Transport of Suspended Substances in Porous Media

This book offers a comprehensive and accessible exploration of the transport of suspended substances in porous media: a process integral to environmental protection, energy development, and geotechnical engineering applications.

Delving into the movement, attachment, detachment, and interactions of particles and dissolved substances under seepage conditions, the volume combines laboratory experiments, analytical modeling, and theoretical advancements. It systematically investigates transport processes involving particles and ions in both saturated and unsaturated porous media, with a particular focus on temperature effects, multi-substance interactions, and nonequilibrium behaviors. A key highlight of this work is the development of a nonlinear attachment–detachment model with hysteresis, capable of describing transport phenomena across a range of particle scales, from ionic to microscale. Furthermore, the book introduces an innovative granular thermodynamic framework that links energy dissipation, seepage, and substance transport, providing a fresh perspective on coupled flow processes. The practical significance of these concepts is illustrated through detailed case studies, including heavy metal migration, groundwater remediation, landfill leachate control, geothermal resource development, and industrial pollutant treatment.

As the first international reference dedicated specifically to the transport of suspended substances in porous media, this volume not only opens new avenues for research but also serves as an invaluable resource for researchers, engineers, and advanced students in geotechnical engineering, environmental engineering, and energy-related disciplines.

Bing Bai is Professor at Beijing Jiaotong University, China. His research focuses on contaminant transport in porous media, granular thermodynamics, and geo-environmental engineering. He has published extensively on coupled transport processes and nonlinear attachment–detachment modeling.

Transport of Suspended Substances in Porous Media

Experiment, Theory and Engineering Implication

Bing Bai

CRC Press
Taylor & Francis Group
Boca Raton London New York

CRC Press is an imprint of the
Taylor & Francis Group, an **informa** business

First edition published 2027
by CRC Press
2385 NW Executive Center Drive, Suite 320, Boca Raton FL 33431

and by CRC Press
4 Park Square, Milton Park, Abingdon, Oxon, OX14 4RN

CRC Press is an imprint of Taylor & Francis Group, LLC

ISBN: 978-1-041-32637-3 (hbk)
ISBN: 978-1-041-32639-7 (pbk)
ISBN: 978-1-003-78374-9 (ebk)

DOI: 10.1201/9781003783749

Typeset in Minion
by Apex CoVantage, LLC

Contents

Preface

The coupled flow of multiphase suspended substances in porous media involves irreversible nonequilibrium processes. The research of the physical essence of multi-substance transport under seepage helps to promote the development of contaminant transport theory, deepen the understanding of the granular thermodynamics, and has important application value in industrial pollutant treatment, groundwater extraction and recharge, geothermal resource development, thermal energy storage, waste disposal in metal mining, and the development of barrier and purification technology for leachate from landfills in shallow infiltration. It meets the major strategic needs of global ecological protection and environmental governance.

For over 20 years, the author of this book has been dedicated to researching the transport of suspended substances in porous media, and has achieved a series of original achievements, including some important scientific discoveries:

(1) The physical nature of coupled flow of multi-suspended substances (such as silica powders, natural shell powders, red mud particles, graphene oxide particles, or heavy metal ions) in porous media is conducted systematically using a self-developed temperature-controlled seepage transport test device. Moreover, the physical mechanism of temperature effect has been revealed for the first time internationally, indicating that it is related not only to the dielectric properties of porous media matrices and suspended substances, but also to the significant differences in attachment/detachment processes and phase transition processes caused by complex physicochemical interactions between these substances.

(2) A nonlinear static attachment–detachment model for suspended substances considering attachment-history and temperature effects is constructed, and an adsorption-kinetics model reflecting the "adsorption lag" effect between suspended substances and porous media solid matrix is established. This model utilizes the constructed "scanning detachment isotherms" to describe the attachment–detachment–reattachment process caused by the increase, decrease, and increase again of suspended substance concentration, which is an important innovation from the previous theoretical models based on the adsorption-kinetics.

(3) Based on the granular thermodynamic principles, the concepts of granular temperature, granular entropy, and granular motion equation are creatively introduced

to reveal the nonequilibrium hydrodynamic processes of attachment, detachment, and transport under the changes in suspended substance concentration and seepage velocity. The author has established a comprehensive control equation for the coupled transport process of multiphase suspended substances and a new calibration method for transport parameters, which can accurately predict the evolution process of coupled transport under engineering backgrounds such as repeated changes in contaminant concentration and increasing or decreasing seepage velocity.

This book consists of seven chapters in total and discusses the relevant issues of the transport of suspended substances in porous media from laboratory tests, analytical solutions, and granular thermodynamic theory. Chapter 1 discusses the concept, significance, and current development status of the transport of suspended substances in porous media. Chapter 2 elaborates the typical experiments and some interesting phenomena in the transport of suspended substances. Chapter 3 mainly proposes a nonlinear attachment–detachment model with hysteresis for the transport of suspended substances as well as the competitive adsorption mechanisms of various suspended substances. Chapter 4 elaborates the typical experiments and physical mechanism on the cotransport of multi-substances under seepage conditions. Chapter 5 discusses the transport processes of moisture and the heavy metal ions in unsaturated soils. Chapter 6 derives the analytical solutions of suspended substance transport by using the source-function method as well as a cotransport theoretical model of multi-substances. Chapter 7 creatively establishes a granular thermodynamic multi-substance flow model on the basis of the energy dissipation caused by the consolidation deformation of porous media under external force and the transport of internal suspended substances suitable for saturated/unsaturated porous media.

This book is the first international monograph on the transport of suspended substances in porous media, which can be used as a reference for researchers and relevant professional teachers in the fields of porous media, geomechanics, environmental engineering, and thermodynamic engineering technology, including theoretical research, design, and construction.

The research results have received strong support from the National Natural Science Foundation of China (52378321; 52079003; 51678043; 51279002; 50978022) and the Key Research and Development Projects of the Xizang Autonomous Region (XZ202501ZY0125). My postgraduate students, including Chen Peipei, Bai Fan, Zhou Rui, Zhang Bixia, Wu Haiyan, Chen Jing, Cui Yiyi, have carried out many effective works for the publication of this book.

CHAPTER 1

Introduction

1.1 RESEARCH SIGNIFICANCE

The transport of suspended substances in porous media, such as soluble ions, heavy metal ions (HMs), bacteria, silicon powders, and red mud particles (RMs), is an important challenging topic in environmental geotechnics and civil engineering, as well as in many other fields, such as groundwater exploitation, water purification, variation in the permeability of soils, contaminant treatment, oil resource exploration, groundwater recharge, thermal energy storage, geothermal resource development, waste disposal in metal mines and landfill leachate barrier generation, and the cotransport of various pollutants.

The interactions between different suspended substances, as well as various treatment methods (e.g., chemical precipitation, reverse osmosis, and electrocoagulation) have been closely studied. Furthermore, some researchers have attempted to utilize various adsorbents, such as natural inorganic diatomite, modified clay mineral particles, biochar, nanomaterials, and activated carbon, to purify water-bearing strata polluted by various traditional and new types of pollutants, which is a significant objective in environmental engineering.

1.2 CURRENT RESEARCH REVIEW

1.2.1 Transport of Suspended Substances

Suspended substances in groundwater are mainly composed of macromolecular organic substances like biopolymers secreted by microorganisms and human excreta, microemulsions, rock fragments, and mineral deposits (e.g., hydroxides, silicates, phosphates, carbonates, manganese, calcium and aluminum oxides, and actinide hydroxides and oxides), with particle sizes ranging from several nanometers to several millimeters (Bai et al., 2025a). Generally, the mobile solid particles with a diameter larger than 10 μm are defined as suspended particles (SPs), while the mobile particles with a diameter less than 10 μm are defined as colloids. Nevertheless, in some other studies, the threshold value between SPs and colloids is set at 1 μm. Actually, the particle size between 1 μm and 50 μm can be considered as a transition from colloids to SPs and the seepage transport of particles in this size range is impacted by physical mechanisms, such as the electrostatic force, hydrodynamic

DOI: 10.1201/9781003783749-1

dispersion, molecular diffusion, particle concentration, gravity, and even size exclusion effect (Bai et al., 2016).

Water quality issues associated with microbial pathogen-induced groundwater pollution, colloid-facilitated contaminant transport, wastewater treatment by packed bed filters, and bioremediation of contaminated groundwater aquifers have stimulated considerable research efforts toward the filtration of colloidal particles (Natarajan and Kumar, 2011). The suspended substances have a high capacity to absorb pollutants and move with seepage water (Sun et al., 2010), and they can also facilitate the transport of contaminants and act as a third phase except the mobile aqueous phase as well as the immobile solid phase of porous media.

Many factors can affect the transport of suspended substances in porous media and their interactions with the solid matrix, including the solution chemistry characterized by solution ionic strength and pH value, double-layer repulsive force, London–van der Waals attractive force, critical total ionic strength, and even ambient temperature, which conversely depend on the pore structure and the size. Obviously, temperature plays a crucial role in the viscosity and aggregation properties of suspended substances (Bradford et al., 2009; Arab et al., 2014). For instance, Tomita et al. (1995) observed that as temperature increased, the amount of ammonium polyacrylate–based dispersant required to achieve the minimum viscosity increased. However, at the optimum dispersant concentration, the minimum viscosity decreased with higher temperatures. This phenomenon was attributed to the greater amount of dispersant adsorbed onto the particle surface as the temperature rose. García et al. (2017) investigated the effect of temperature on the aggregation kinetics of colloidal montmorillonite particles in aqueous dispersions and found that the rate constant for colloid aggregation increased with increasing temperature when $pH < 4$, but decreased with increasing temperature when $pH > 10$. This effect became more pronounced at higher ionic strengths and temperatures. Tari et al. (2000) showed that increasing temperature led to a gradual diminution of alumina surface ionization, dielectric constant, and a total energy barrier for the coagulation, and the enhancement of the coagulation rate with increasing temperature led to a gradual increase of both relative apparent viscosity and thixotropy.

Temperature significantly affects the transport characteristics of suspended substances. Schembre and Kovscek (2005) reported a temperature-dependent permeability reduction in Berea sandstones during hot-water flooding experiments (e.g., 20–200 °C), with fine particle mobilization occurring at distinct temperature thresholds influenced by solution pH value and ionic strength. Subsequently, Kim and Walker (2009) investigated the effect of temperature on the transport of two bacteria (*Escherichia coli* strains) and a model-colloid (surfactant-free carboxylate-modified latex, 1 μm) using packed-bed column experiments. These results showed that the deposition coefficient for the colloid at 25 °C was approximately three times greater than that at lower temperatures (4 °C and 10 °C). Moreover, colloids exhibited higher deposition rates than bacteria at elevated temperatures; nevertheless, the reverse trend was observed under cooler conditions, highlighting the differential sensitivity of SPs to thermal variations. Furthermore, Rosenbrand et al. (2012) verified that kaolinite particles had a heterogeneous surface charge distribution, which plays a significant role in the particle aggregation in the pore throats of the sandstones and thus leads to permeability damage. Additionally, the reduction of temperature induced an increase in the hydrodynamic force on the aggregated particles in the pore throats. These observations can be interpreted qualitatively by the classical Derjaguin–Landau–Verwey–Overbeek (i.e., DLVO) theory (Rosenbrand et al., 2012).

Actually, the rise of temperature changes the adsorption properties between suspended substances and between the suspended substances and the matrix of porous medium, and enhances the intrinsic motion energy of suspensions, which eventually influences the movement of suspended substances. Temperature also has an obvious effect on the attachment stability of suspended substances. Previously, Seebergh and Berg (1995) elaborated the surface properties of amidine and sulfate latexes and found that increasing the temperature changed the equilibrium state between the ionosphere and porous media, which altered the surface potential of suspended substances and increased the repulsion potential between the suspended substances and the porous media. That is, deposition due to the effect of gravity decreased (Chrysikopoulos and Katzourakis, 2015). Sandra et al. (2006) measured the surface charge (ζ-potentiodynamic potential) during bentonite particle deposition at different temperatures (0, 20 °C, and 80 °C) and demonstrated that the particle stability increased with temperature due to the temperature dependence of the surface charge. Similarly, Hyunjung and Sharon (2009) pointed out that the adsorption coefficient of carboxylic ester emulsion-modified colloids (*carboxylate-modified latex*, particle diameter of 1 μm) at 25 °C was three times higher than that at 10 °C. This result can be attributed to the change in surface potential between suspended substances and the solid matrix, leading to increased movement and a more active thermodynamic collision process (Alem et al., 2015; Monged et al., 2018). An increase in temperature also reduced the surface charge density of suspended substances. The thickness of the electric double layer (i.e., adsorption and diffusion layers) and the stability of the suspended substances and solid matrix interface decreased according to the DLVO theory (Tari et al., 2000). Temperature effect gradually increased with decreasing suspended substance particle size (Sandra et al., 2006). A small number of pollutant transport experiments considering the temperature effect have been conducted (Bai et al., 2017a), and they showed that temperature changed the adsorption-kinetics of pollutants in porous media due to variations in the electrical double-layer thickness.

Straining is a key factor affecting the transport of suspended substances (Chrysikopoulos and Syngouna, 2014; Liu et al., 2016). Generally, the finite size of a colloid particle excludes it from the slowest moving portion of the parabolic velocity profile within a fracture or a pore throat (Bennacer et al., 2013), thus the effective particle velocity is increased (Ovaysi and Piri, 2011); the overall particle dispersion is reduced, but with a tendency to increase with increasing particle size (Bennacer et al., 2013; Bai et al., 2017b). However, inconsistent results have also been reported (Keller et al., 2004). Some studies have shown that there existed a critical ratio of the particle diameter to the collector diameter, below which the straining rate was negligibly low (Vasiliadou and Chrysikopoulos, 2011).

The influence of pH value on the transport of suspended substances is multifaceted, primarily manifested in variations in particle adsorption/desorption processes and permeability (James and Chrysikopoulos, 2011). Kim and Walker (2009) analyzed the effect of pH on the aggregation as well as deposition of polymer coating of nanoscale zero-valent iron on the surface of different porous matrices of soils. Wang et al. (2010) discussed the adsorption of tetracycline onto the clay and marine sediment from seawater at different pH levels and ionic strengths, and the results indicated that the adsorption quantity decreased with increasing pH and ionic strength. The research of Kobayashi et al. (2013) in a packed bed of spherical zirconia beads indicated that the maximum coverage of sulfate latex with a diameter of 2 μm on the matrix surface was constant irrespective of pH, whereas that of carboxy latex with a diameter of 0.51 μm decreased with increasing pH. Tusara et al. (2015) experimentally observed that the presence of suspended materials (<0.45 μm) caused a significant decrease in permeability in the porous column, but the decreasing rate could be reduced by lowering the pH value. Wang et al. (2015) indicated that humic acid (HA—an organic colloid matter) had a less significant effect on arsenic (As) transport in water-saturated sand columns at high pH (e.g., pH = 9) than at low pH (e.g., pH = 5) due to the low adsorption of HA. Haliena et al. (2016) found that the enhanced transport of thorium(IV)—a representative actinide—in sand columns was primarily due to the pH-dependent desorption of Th(IV) caused by the change in ionic strength.

Changes in pH can affect the interface properties between particles and between the particles and the matrix of the porous medium due to the electric double-layer action, which in turn can affect the adsorption properties of suspended substances. This effect also depends on the possible chemical reactions of various suspended substances and matrix materials with alkalinity in addition to their physical properties (Fu et al., 2018). Wang et al. (2016) revealed that silica colloid with a diameter of 0.45 mm was more likely to deposit on the surface of the porous media under acidic conditions (pH = 5) than under alkaline conditions (pH = 9), indicating a high pollution risk for the recharge water with high pH. However, despite the recognized importance of alkalinity in the transport of colloidal particles (<1 μm) due to their abundance in groundwater and physical properties, such as specific surface area and large attachment-capacity (Bradford et al., 2009; Jaiswal et al., 2018), little research has focused on the transport of suspended substances with the size larger than 1 μm at different alkalinities in saturated porous media (Arab et al., 2014; Ahfir et al., 2017).

The classical Langmuir or Freundlich static adsorption models can be fitted to results showing the attachment performance in water (García et al., 2017). The main factors affecting the attachment of ions (or suspended substances) to porous media include the pore structures, the types of pollutants, and the chemical environments of the aqueous solutions (Bradford et al., 2017). In particular, suspended substances ranging from 0.1 μm to 50 μm in size are subject to a range of transport forces, including molecular diffusion, electrostatic interactions, thermodynamic gradients, hydrodynamic drag, and gravitational settling (Bedrikovetsky et al., 2012; Chrysikopoulos and Syngouna, 2014). The DLVO theory provides a foundational framework to assess the total surface interaction potential between particles and solid surfaces (Feng et al., 2021). However, the interference of chemical factors (Bedrikovetsky et al., 2017), such as a decrease in the ionic strength (Nishad and Al-Raoush, 2021) or an increase in the pH, can significantly change the contact force between multiphase materials (Wyness et al., 2018). Empirical studies (Sotirelis and Chrysikopoulos, 2017) applied pseudo-second-order kinetic models to describe the pH value and ionic-strength-dependent attachment of graphene oxide (GO) to quartz sand and kaolinite colloids (Bradford et al., 2017; Syngouna et al., 2020).

The essentially nonlinear attachment and detachment processes involve complex physicochemical effects (Daghooghi and Borazjani, 2018; Turkyilmazoglu, 2020) caused by spatiotemporal variations in the particle size (Bai et al., 2017a; Gavrilov and Shebelev, 2018) and concentration (Russell and Bedrikovetsky, 2018). These complex processes involve physicochemical interactions, often deviating from first-order kinetic assumptions that dominate conventional modeling approaches (Katzourakis and Chrysikopoulos, 2018). Importantly, few models incorporate the effects of previous deposition processes (so-called attachment-history). Farajzadeh et al. (2016) coupled an established transport theory with the static Langmuir isothermal equilibrium adsorption model and considered two physical mechanisms, namely, physical adsorption and mechanical capture, to explain the attachment process of injected organic polymer particles to understand the size effect of these particles and the changes in the permeability and pore pressure of the porous media over time. Bedrikovetsky et al. (2017) studied the adsorption of modified single-size carboxyl latex particles (carboxyl groups) in porous media with a random pore size distribution, focusing on modifying the attachment/detachment model to account for the effect of pore structure evolution.

1.2.2 Cotransport of Multi-Substances

Contaminants in porous medium can be roughly classified into three types: contaminants adsorbed on immobile solid particles, contaminants dissolved in water, and contaminants adsorbed on the surface of solid particles that move with water flow (Chrysikopoulos and Katzourakis, 2015). Oxides, clay minerals, and colloidal humus particles that exist ubiquitously in the underground environment may accelerate the transport of contaminants due to their large specific surface area and high sensitivity to the physical and chemical environment (Ahfir et al., 2017). Silicon powders play an important role in soil and aquifer contamination and those suspended substances transported easily in the water flow can act as carriers of contaminants (Haliena et al., 2016; Wang et al., 2016). However, suspended substances can also be a barrier to the transport of contaminants if their presence clogs the porous medium (Li and Zhou, 2010; Bennacer et al., 2017).

Colloids are recognized as important facilitators of heavy metal transport (Kim and Walker, 2009; Natarajan and Kumar, 2011). For example, Puls and Powell (1992) revealed that the transport rate of colloid-associated arsenate (As) was more than 21 times that of dissolved As. Kersting et al. (1999) found that the radionuclides in the aquifers at the Nevada test site, where hundreds of underground nuclear tests were conducted, were associated with the colloidal component of the groundwater, and mobile colloids had the potential to enhance the transport of nonsoluble contaminants due to the adsorption effect (Wang et al., 2000). Similarly, Karathanasis (1999) also showed that the presence of colloids obviously enhanced HMs (e.g., zinc) transport by 5- to 50-fold over that without colloids, with Zn^{2+} being consistently more mobile than Cu^{2+}. Yin et al. (2010) investigated colloid-facilitated lead transport in shooting range soils and showed that there was a significant correlation between colloids and Pb^{2+} in the leachates, and both the decrease in ionic strength and the increase in flow rate contributed to the release of colloids and Pb^{2+}.

Generally, colloids provide a pathway for the rapid transport of HMs like Pb^{2+} (Grolimund et al., 1996). Pang et al. (2005) studied the transport of bacteria-facilitated cadmium by column experiments in a saturated gravel aquifer and analyzed the influences of many factors such as bacteria, solution pH, surface charge, and bacterial size. Missana et al. (2008) investigated the coupled effect of the presence of bentonite colloids ranging from 0.2 μm to 0.25 μm in size on the seepage transport of elements such as Eu and Pu in a granite environment, and found that the seepage transport of these radionuclides was mostly colloid-driven. Wang et al. (2015) showed that illite colloids (<0.1 μm in size) did not influence As transport in water-saturated sand columns due to their low affinity for As, and humic acid substantially increased As transport via competition against As for adsorptive sites on the transport medium and potential formation of nonadsorbing aqueous phase As–humic acid complexes. Thus, it is concluded that clay–mineral colloids in general may not have a large potential to increase As transport. Ma et al. (2016) investigated the cotransport of soil colloids (<1.2 μm in size) and As at different pH and ionic strength in saturated sand columns, and the results showed that the strong repulsion between the porous media and the soil colloids promoted As transport by blocking the adsorption of As onto the porous media, and the colloids acted not only as a carrier in facilitating solute transport but also as a barrier in the course of solute adsorption. Zhou et al. (2016) studied the cotransport of a single layer of graphene oxide particles with a diameter of 0.7811±0.5022 μm and a thickness of 0.6±1 nm and Cu^{2+} in granular natural sand columns, and concluded that GO had fairly high mobility and could serve as an effective carrier of Cu^{2+}.

The transport of HMs is also governed by interactions with soil matrices and environmental conditions (Bai et al., 2020a). Besides, Wu and Li (1998) reported that the mobilities of several typical HMs followed the order of $Cu^{2+} > Zn^{2+} > Pb^{2+} > Cd^{2+}$ in neutral leachate (pH = 7) but followed the order of $Cd^{2+} > Zn^{2+} > Cu^{2+} > Pb^{2+}$ in more acidic leachate (Valsala and Govindarajan, 2019). Tang et al. (2013) indicated that water-soluble organic carbon could compete for adsorption sites or form soluble complexes with HMs to reduce their adsorption onto soil media to facilitate the transport of HMs. Kumpiene (2015) investigated the microbial functional diversity, biochemical activity, HMs availability, and soil toxicity of Cd^{2+}-, Pb^{2+}-, and Zn^{2+}- contaminated soils to restore soil ecological functions. Ma et al. (2016) elaborated that the rapid transport of soil colloids facilitated As(V) transport by inhibiting As(V) adsorption onto sand. Lee et al. (2019) revealed that the outflow concentration of HMs, such as Cr^{2+} and Pb^{2+}, in the pore solution increases with increasing colloidal particle concentration, while Cd^{2+} and Cu^{2+} lagged behind the release of colloidal particles; that is, the physicochemical effects in the solution notably impacted the HMs transport promoted by different SPs.

Building on these findings, further studies have explored the combined effects of mineral composition, solution chemistry, and particle characteristics on the coupled transport behaviors of HMs and SPs in porous media. Kumpiene et al. (2008) reviewed the differences in the adsorption processes of HMs (including Cu, Cr, Pb, As, and Zn) on soil particles as well as the chemical reactions and physical mechanisms underlying the transport processes, and suggested corresponding stabilization mechanisms and technological measures for heavy metal pollutants. Bai et al. (2016) found that there exists a threshold value for the size of SPs; nevertheless, when the size of SPs is smaller than the threshold, the promoting effect of SPs on the transport of HMs increases with increasing particle size; otherwise, its promoting effect decreases significantly with increasing particle size. Du et al. (2018) found that kaolinite has a much higher attachment-capacity for Pb^{2+} than sand; saturated sand column experiments showed that the existence of kaolinite particles enhances the transport of Pb^{2+} and that the presence of Pb^{2+} in turn reduces the transport of kaolinite particles (Bai et al., 2020c). By reducing the HMs concentration and varying the chemical state of the seepage water flow, Zhu et al. (2014) studied the transport process and release effect of Hg ions in contaminated soils; in this case, the organic matter acted as a carrier for Hg transport in soils. The pH also affects the attachment-capacity of soil for HMs (Yu et al., 2019). For example, the adsorption rate of Pb^{2+} in soils increases with increasing pH at pH <6.5, whereas it decreases with increasing pH at pH >8. Nishad and Al-Raoush (2021) conducted experiments using a hydrophilic glass micromodel to examine the influence of the solution ionic strength, hydrophobicity, and pH on the colloid retention in the case of multiphase flow.

Ahfir et al. (2017) investigated the removal performance of suspended substances with a diameter range of 2–30 μm in porous media based on median diameter variations of the deposited suspended substances along a column and breakthrough curve evolution in the effluent. Li et al. (2018) simulated the microscale changes in colloidal particle transport in porous media and replicated the flow field by reconstructing the pore structure, which focuses on the deposition probability and the spatial distribution of deposited particles. The deposition process is always inseparable from the detachment mechanism. In this aspect, Cui et al. (2017) aimed to delineate the detachment characteristics of deposited quartz powder particles, and they revealed that changing the flow direction was more effective than changing the flow rate through a two-dimensional sandbox packed with quartz sand. Despite the many studies on the coupled transport of HMs and SPs in flowing water (Bradford et al., 2009; Katzourakis and Chrysikopoulos, 2024), the deposition mechanisms of HMs together with SPs along the transport distance remain to be elucidated (Kim and Walker, 2009).

As a method of removing heavy metal pollutants from groundwater or wastewater, the adsorption treatment has the advantages of high efficiency (Sotirelis and Chrysikopoulos, 2017), no secondary pollution, simple operation, and low energy consumption, and it has shown favorable application prospects (Xu et al., 2022). Activated carbon, zeolite, biomaterials, and clay are common adsorbent materials (Deliyanni et al., 2015), and the evolutionary characteristics of their attachment–detachment properties are important indicators for choosing the absorbent for the specific situation (Saavedra et al., 2019; Cui et al., 2022). The desorption characteristics of adsorbed pollutants are an important aspect of the performance of adsorbents, and they also involve related issues such as the solidification performance of heavy metal pollutants (Luo et al., 2021; Zhu et al., 2022) and the reliability of soil remediation. Peld et al. (2004) looked into the attachment–detachment characteristics of Zn^{2+} and Cd^{2+} on apatite and found that the solute concentrations of NO_3^- and Cl^- had a minor influence on adsorption behavior but a great influence on desorption behavior. Shirvani et al. (2006) discussed the nonsingularity of the isothermal-adsorption curves of Cd^{2+} in palygorskite and sepiolite as well as in calcite systems, finding that the isotherms of adsorption/desorption reactions were not completely consistent; obvious hysteresis was present in the adsorption and desorption processes of Cd^{2+}. Covelo et al. (2007) discussed the adsorption characteristics of 6 heavy metal pollutants in 11 acidic soils. According to the partition coefficients, the adsorption of Cu^{2+} and Pb^{2+} was preferred over that of Cd^{2+}, Ni^{2+}, and Zn^{2+}, and the clay minerals in the soil (such as kaolinite, gibbsite, and vermiculite) played a key role. Milosevic et al. (2021) analyzed the desorption efficiency of a novel material (i.e., *Handkea utriformis*) in wastewater treatment, showing that hybrid subgleba-based adsorbents are obvious choices for the active remediation of leachate polluted by HMs.

In addition to a large number of soil column experiments, many studies have focused on attachment–detachment models as well as the establishment and solution of the governing equations for the coupled transport of various types of suspended matter (Bai et al., 2015), with satisfactory theoretical prediction results (Lindner et al., 2017). Bedrikovetsky et al. (2017) proposed a method for determining the pore size distribution of a porous medium from a long-term monosized suspension injection and a theoretical model for predicting the transport process. This model avoids the drawbacks associated with the analysis of the results from short-term injection tests using multiple-size particle suspensions and considers the constraining effect of the pore scale of the porous medium on the particles in the water and the nonlinear characteristics caused by their cumulative deposition process.

In fact, the factors that affect the coupled transport of SPs and HMs include not only the size of the SPs (Turkyilmazoglu, 2020; Chequer et al., 2021), concentration of the particle solution, seepage velocity, temperature effect (Bai et al., 2017a), type of HMs, and pH (Hudcova et al., 2018; Nishad and Al-Raoush, 2021), but also factors such as the type and physiochemical properties of the porous media and the coupling between different types of multiphase multicomponent substances (e.g., solid–liquid, solid–solid, and liquid–gas) (Chequer et al., 2020), and even the chemical reactions between them (Nishad and Al-Raoush, 2021). Using special barrier materials, Abbar et al. (2019) deliberated the cotransport of SPs and HMs (including Cu^{2+}, Pb^{2+}, and Zn^{2+}) in sand columns with and without flax geotextiles; this method improved the sand retention feasibility by retaining the soluble and attached HMs in the pore structures (Turkyilmazoglu, 2019; Sharma et al., 2020).

1.2.3 Transport of Suspended Substances in Unsaturated Soils

The theory established by Philip and De Vries (1957) is known as a classical frame that is a comprehensive and representative theory of moisture/heat transport in unsaturated porous media. This formulation has attracted considerable interest that has led to a number of publications. Initially, it is assumed that liquid water, water vapor, and heat transfer occur simultaneously, while each flow is independently related to the gradients of temperature and moisture content. Thomas and King (1992) presented a theoretical discussion on the heat and moisture movement in unsaturated soil based on the assumption that the total potential for liquid flow consists of two components, the elevation and the capillary potential; therefore, the fundamental potentials employed in the work are temperature and capillary potential. Basha and Selvadurai (1998) developed an analytical approach to the problem of heat-induced moisture movement in the vicinity of a spherical heat source embedded in an undeformable, moist, porous, infinite solid and illustrated the influence of moisture movement on the time-dependent distribution of moisture. Thomas et al. (1998) established a model for the transport of heat, moisture, and air, which is applicable to deformable unsaturated soil.

Zhang et al. (2015) developed a multiscale theoretical model for investigating heat and mass transfer between unsaturated sandy soils and biochar, which can evaluate the effects of biochar on sandy soil water retention. Varnosfaderani et al. (2016) established a fully coupled thermo-hydro-mechanical (THM) model for describing the non-isothermal evolution of an unsaturated porous medium, in which a new thermal state surface of saturation degree was introduced to couple moisture content with temperature. Hassini et al. (2017) developed a two-dimensional drying model that coupled all heat and water transfer mechanisms (advection, filtration, and capillarity), which was applied to the convective drying of unsaturated and viscoelastic materials (e.g., clay–cellulose fiber composites). Previous research assumed that the thermo-hydraulic properties are constant and that the heat transfer in soil occurs only in the form of conduction, neglecting convective and latent heat transfer. Moradi et al. (2016) evaluated the significance of variable thermal and hydraulic properties and different heat transfer mechanisms in unsaturated soils. Jahangir et al. (2018) presented a numerical model for simulating coupled heat and mass transfer in unsaturated soils; the proposed model was applied as a thermal storage system to measure the thermal behavior by changing every relevant parameter.

Heat transfer leads to significant changes in the temperature potential and suction potential in unsaturated porous media and causes the moisture to move from high-temperature area to low-temperature area. This movement is accompanied by the transport of contaminants through porous media; common contaminants include organic pollutants, inorganic pollutants, and even HMs. In the past, studies on the transport process and retardation mechanism of pollutants were generally conducted under the driving force of a hydraulic gradient and a chemical gradient, whereas the transport effect caused by a thermal driving force was not elucidated. Funk and Wakili (2008) established a linear transport process model describing matter transport in porous media caused by thermal driving forces, wherein this model accounted for the effects of moisture content, vapor pressure, and chemical potential. Later, Miracapillo and Ferroni (2015) simulated the transport of radionuclides in the far field of a geological repository in a nonuniform groundwater flow and investigated the effects of the distribution coefficient. Tang and Cui (2010) investigated the influences of suction as well as temperature on the compression of the unsaturated compacted MX80 bentonite and reported the transport of radioactive pollutants.

On the other hand, the smoothed particle hydrodynamics (SPH) is a Lagrangian mesh-free particle method consisting of two steps of approximation: the kernel approximation and the particle approximation. In the kernel approximation, the function and its derivatives are represented as integrations over the support domain; while in the particle approximation, the support domain is divided into discrete particles with their own masses, densities, velocities and pressures. The SPH was initially developed to solve astrophysical problems in unbounded domains, in which the governing equation could be expressed by the classical Newtonian fluid dynamics. Subsequently, this method was also successfully applied in fluid dynamics and free surface flows and non-Newtonian multiphase flows.

A major concern in the application of SPH method is the physical type of boundary conditions (BCs), such as the solid BC, open BC, inflow/outflow BC, and periodic condition. Fang et al. (2006) simulated transient viscoelastic free surface flows and introduced the artificial pressure between particles to stabilize the SPH system avoiding the tensile instability. Ataie-Ashtiani et al. (2008) proposed an incompressible SPH method to simulate free-surface incompressible fluid problems. Liu and Liu (2010) proposed an improved SPH method for resolving discontinuous interfaces of shock waves in aerodynamics based on the Taylor series expansion. Fourtakas et al. (2015) presented a method to impose 2D solid-wall boundary conditions in SPH, by which arbitrary complex domains could be readily discretized ensuring approximate zeroth- and first-order consistency. These findings provide an effective approach for the calculation of SPH boundary problems. The SPH method is well-suited for thermal processes of some complicated surface flows due to its Lagrangian nature (Cleary and Monaghan, 1999). Jeong et al. (2003) developed an algorithm for the boundary condition implementation by decomposing a second-order partial differential equation (PDE) into two first-order PDEs, which was applicable to complex geometries and nanoscale heat transfer. Schwaiger (2008) established a SPH formulation of the Laplacian operator that could greatly improve the accuracy near free boundaries based on a gradient approximation commonly used in thermal problems. In addition, Alshaer et al. (2017) investigated the SPH modeling of transient heat transfer in pulsed-laser ablation of aluminum where the laser was applied directly to the free-surface boundary of a specimen.

The mechanism of the transport of soluble pollutants is via the redistribution of moisture in the pores as a carrier in soils. Seepage flow in porous media plays a leading role and is accompanied by the transport and adsorption of heavy metal pollutants (Cui et al., 2017; Ziedan et al., 2024). In fact, the transport mechanism of HMs includes multiple effects, such as convection, molecular dispersion, and mechanical dispersion (Bai et al., 2021a; Feng et al., 2021). Whether in unsaturated/saturated soils, the molecular diffusion coefficient of pollutants is closely related to the water content (Markale et al., 2022). If the seepage flow in the pores is quick, the effect of mechanical dispersion is greater than that of molecular dispersion (Gong et al., 2021), and the effect of molecular diffusion can be neglected. Conversely, if the seepage velocity of pore water is small, only the effect of molecular dispersion can be considered (Muraleedharan and Nadir, 2021). In addition, the transport of HMs is also affected by the physical characteristics of soil particles, pollutant characteristics, and external environmental factors, which deserve attention (Bennacer et al., 2017; Nishad and Al-Raoush, 2021).

1.2.4 Theoretical Methods of Suspended Substance Transport

Some mathematical models have been proposed to characterize the colloid-facilitated transport of contaminants in the dissolved phase (Sen and Khilar, 2006), which generally comprise the mass balance equations for solid particles and contaminants in different forms and the reaction processes among different constituents (Corapcioglu and Jiang, 1993). Bekhit et al. (2009) studied the combined effect of colloids and bacteria on the contaminant transport and then proposed a conceptual model considering multiple reactions, which was suitable for the case of three mobile constituents and a fixed solid matrix phase. Katzourakis and Chrysikopoulos (2014) also developed a mathematical model for the simultaneous transport of viruses and colloids in saturated homogeneous porous media with uniform flow. Certainly, the feasibility of these mathematical models relies heavily on a good understanding of the coupled transport of multiple mobile and/or immobile species in aqueous phase.

Chang et al. (2001) simulated Cu^{2+} and Pd^{2+} transport in a lateritic silty-clay soil column by using the Freundlich equilibrium-controlled adsorption parameters to determine the retardation factor used in column leaching experiments. Peters and Smith (2002) derived flow and transport equations for a deforming porous medium based on the mass balance law and illustrated the differences between the theory for the rigid porous medium and deforming porous medium using solute transport through an engineered landfill liner as an example. Jungnickel et al. (2004) described the solution of a fully coupled set of transport equations describing the simultaneous diffusion of several ion species through a clayey soil. Seetharam et al. (2011) derived a multicomponent reactive transport model, coupled with an existing thermal, hydraulic, and mechanical model for porous media, based on conservation of mass/energy principles for flow and stress–strain equilibrium. Considerable effort has also been devoted to the modeling of consolidation-induced contaminant transport. Fox et al. (2011) computationally investigated the importance of the consolidation process to solute transport in compressible porous media.

Analytical solutions for more complex problems, such as variable coefficients, unsteady flow, and multilayered media, have gained attention. Kumar et al. (2010) used the Laplace transform to derive the analytical solutions for the one-dimensional advection–diffusion equation, with variable coefficients and initial and boundary conditions in a semi-infinite domain for three dispersion problems; the solutions could be used for a variety of dispersion and velocity variations due to their spatial and temporal dependence. Guerrero and Skaggs (2010) presented a general analytical solution for a one-dimensional advection–dispersion equation with distance-dependent coefficients, deduced a transport equation with a self-adjoint differential operator by using an integrating factor, and obtained the solution using the generalized integral transform technique. Li and Cleall (2010) established analytical solutions for conservative solute diffusion in one-dimensional double-layered porous media; these solutions are applicable to various combinations of fixed solute concentrations and zero-flux boundary conditions applied at each end of a finite one-dimensional domain and can consider arbitrary initial solute concentration distributions throughout the media. Chen et al. (2011) presented exact analytical solutions to the two-dimensional advection–dispersion equation in cylindrical coordinates in a finite domain subject to the first-type and third-type inlet boundary conditions (i.e., the Dirichlet boundary condition and mixed boundary condition) using the finite Hankel transform of the second kind and the generalized integral transform technique. Yadav et al. (2011) discussed the analytical solution of a one-dimensional advection–diffusion equation by using the Laplace transform, in which the solute transport and flow field were considered to be unsteady independent patterns.

Several cotransport models for contaminant transport have been proposed (Valsala and Govindarajan, 2019). Bekhit et al. (2009) proposed a theoretical model to account for the transport mechanisms of contaminants, reaction kinetics, and different physicochemical and biological processes in the presence of colloids and bacteria in groundwater, which was verified against experimental data for Cd^{2+} transport in the absence and presence of *E. coli* vegetative cells. Katzourakis and Chrysikopoulos (2014) developed a conceptual model that describes the cotransport of viruses and colloids in a saturated porous medium with uniform flow, which could account for the transport of individual viruses and colloids as well as viruses attached to colloids. Borazjani and Bedrikovetsky (2017) derived analytical solutions for a two-phase flow. It is known that the cotransport of HMs and suspended substances critically depends on their charge characteristics. Positively charged HMs can be adsorbed onto the surface of negatively charged suspended substances (e.g., SiO_2 particles). At the same time, the small-sized suspended substances (e.g., silicon powders) can accelerate the transport of HMs because of the size exclusion effect (Russell and Bedrikovetsky, 2018; Bai et al., 2019a), and yet suspended substances with a size larger than the soil pore dimension can be entrapped in pore constrictions, hence inhibiting the transport of HMs. In addition, charged suspended substances are likely to be adsorbed onto the surface of an oppositely charged solid matrix, resulting in inhibition of the transport of suspended substances. Thus, repulsion between charges can greatly facilitate the transport of suspended substances.

Previously established coupled transport models are predominantly based on convection–diffusion–attachment processes (Kasper et al., 2017; Wang et al., 2023), with an emphasis on attachment–detachment kinetics of the coupled transport process (Katzourakis and Chrysikopoulos, 2014), involving complex physicochemical processes (Li et al., 2017; Yan et al., 2021). The process of attachment/detachment of suspended substances depends not only on the concentration in solution at a given moment, but is also closely associated with the number of deposited suspended substances and changes in matrix structure at that moment (Bedrikovetsky et al., 2017). In other words, the attachment–detachment kinetic process of suspended substances is essentially inseparable from the attachment-history of the porous medium. Ding et al. (2020) adopted the equations of porosity change due to particle deposition and strain, so as to construct a mathematical model of two-phase fluid–solid coupling in the saturated zone of a foundation affected by fine particle transport. In light of the concept of particle rearrangement at meso-level, the establishment of the constitutive model of geotechnical materials based on granular thermodynamics represents a novel approach (Bai et al., 2021d, 2025b). Bai et al. (2023) proposed the concepts of granular temperature, granular entropy, and founded a complete theory of continuous medium mechanics and directly depicted the deformation of materials at meso-level. In particular, the energy dissipation effect caused by seepage velocity has been taken into account in establishing the relationship between dissipative flow and dissipative force. This method circumvents the limitations of the yield criterion, flow rule, and hardening rule, which have a broad range of applicability (Zhou et al., 2024).

Although several studies (including mechanical and hybrid theoretical approaches) have been applied to investigate various THM coupling processes, research incorporating multi-flow and phase-change into the research of soil drying processes remains limited (Bai, 2010; Sendula et al., 2024). Malgaresi et al. (2020) explored the effects of different water-saturation levels on temperature variation, phase-change, and the transport of gases and moisture induced by laser irradiation in porous media, providing valuable insights. Gan et al. (2024) developed a pore-scale lattice Boltzmann model to simulate the coupled heat and mass transfer processes in soils. The established model includes a multicomponent multiphase pseudo-potential and an enthalpy-based phase-change model. Zhou et al. (2024) combined the Van Genuchten soil–water characteristic curve model to develop a theoretical framework for predicting pore solution content and crystallization levels, and analyzed the water–salt phase-change behavior in saline soils through numerical computations. Xiao et al. (2024), based on phase diagram theory, determined the roles of ice and hydrated salts in pore solutions at different concentrations and temperatures. Building on this, they established a multi-physics coupling model that considers ice–salt crystallization, simulating water and salt transport processes as well as the deformation patterns of saline soil during cooling.

Compared to the well-studied saturated media, our understanding of transport processes in unsaturated porous media remains limited (Saeibehrouzi et al., 2024). Zhou et al. (2021) established a mathematical model for the fully coupled processes of heat-water-solute-gas-force under a temperature gradient, based on porous media theory and continuum mechanics principles. This type of model accounts for the influence of factors such as porosity evolution, liquid water flow and its evaporation–condensation processes, gas transport, the dynamic transformation of salt adsorption/precipitation–dissolution, and the compressibility of solid particles and pore fluids. Li et al. (2024) established a coupled mathematical model for water-heat-gas-salt-force in unsaturated sulfate–saline soil, based on unsaturated soil mechanics and porous media theory. This model considers the effects of ice–water phase-change, sodium sulfate crystallization, and vapor flow on heat and mass transfer during the freezing process. Wang et al. (2024) developed a numerical model, considering the THM coupling effect, to describe the seepage and heat conduction in unsaturated soils. The solute transport equation considered convective–dispersive transport in the liquid phase and diffusion in the gas phase (Bai et al., 2020b, 2025c), while the effects of rainfall and evapotranspiration on transport were incorporated into the boundary conditions of the control equations. Shu et al. (2023) discussed the complex interactions between saturated and unsaturated porous media and the transport mechanisms of pollutants in column experiments, focusing on the environmental factors of various pollutants and providing an analytical summary of the underlying mechanisms governing pollutant transport.

1.3 INNOVATIVE WORK OF THIS BOOK

This book elaborates on the transport of suspended substances in porous media from the following aspects:

1.3.1 Chapter 2: Experiments on the Transport of Suspended Substances

The main purpose is to experimentally investigate the effect of particle size, particle concentration, and seepage velocity on the transport of SPs with different particle sizes in porous media. Based on the classical advection–dispersion equation, this chapter discusses the effect of dispersive flux on the deposition dynamics to simulate the breakthrough processes of laboratory column tests.

The chapter investigates the influence of temperature on the seepage transport of SPs, and analyzes the changes of some transport parameters, such as particle transport velocity, attachment coefficient, longitudinal dispersivity, as well as recovery rate.

This chapter also experimentally investigates the effect of alkalinity on the transport of natural silicon dioxide (SiO_2) solid particle suspension injected instantaneously into a saturated porous medium (e.g., quartz sands).

1.3.2 Chapter 3: Attachment–Detachment Model with Adsorption Hysteresis

A nonlinear attachment–detachment model with hysteresis for the transport of suspended substances is proposed. This model uses an attachment-function (or adsorption-function) and scanning detachment isotherms (or desorption isotherms) to simulate the detachment process of suspended substances. The attachment/detachment curves ranged from linear to nonlinear for HMs, GO, and silica powders (i.e., ranging from ions to insoluble particles). The attachment mechanism mainly involved the interfacial electrostatic attraction between suspended substances and the solid matrix, the formation and directional adjustment of metal bonds, and the complementary molecular structure characteristics of different materials. An experiment on the static attachment of HMs (e.g., Pb^{2+}, Cd^{2+} and Cu^{2+}) by RMs was carried out, and the influences of the type of HMs, solution concentration, pH value, reaction time, RMs dosage, and temperature on attachment performance were discussed.

The competitive attachment mechanisms of different coexisting HMs by RMs were compared, which provide a foundation for establishing purification technologies for heavy metal pollutants in water bodies or shallow groundwater. The RMs loaded and not loaded with HMs were detected by zeta potential measurement, scanning electron microscopy, and Fourier transform infrared spectrometry.

1.3.3 Chapter 4: Experiments on the Cotransport of Multi-Substances

The main purpose is to experimentally investigate the transport of HMs in the presence of silicon powders through a sand column under different conditions of seepage velocity, injection HMs concentration, and particle size. The injected SPs possess a very wide particle size distribution (PSD). The concentrations of deposited HMs and SPs and the transport distance along the water flow direction were measured by sampling. The PSDs of the deposited SPs in different sections were obtained by laser diffraction after transport test completion, and microstructure photos were also acquired through metallographic microscopy. The transport of silica powders, which are a type of suspended substance, and HMs under unsteady flow as well as variable HMs injection concentrations, was investigated. During the transport of HMs alone, the HMs concentration in the effluent was almost constant during the entire penetration process, that is, the transport of HMs was not affected by changes in the seepage velocity.

1.3.4 Chapter 5: Transport of Heavy Metal Ions in Unsaturated Soils

A one-dimensional soil column was used to explore the effects of HMs types and concentrations on the soil moisture distribution and the transport level of different HMs, especially during the dynamic change from an unsaturated state to a saturated state. As a special problem, the SPH method was employed to model the heat transfer process in porous media at the pore scale. The effective thermal conductivity can be predicted through a simulation experiment of SPH. The accuracy of the SPH simulation experiment was verified by comparing the predicted values with reference values for ideal homogeneous media and multiphase layered media. The 3D simulation experiments were implemented in granular media generated by the particle flow code (PFC) method.

1.3.5 Chapter 6: Theoretical Methods of Suspended Substance Transport

The general solutions for contaminant transport in a saturated semi-infinite porous media are derived by using the Laplace transform and Fourier transform, along with their transform inversions, under the conditions of seepage flow and the three-dimensional dispersive effect. The analytical expression of contaminant concentration, subjected to a local contaminant source with arbitrary geometry, and intensity that varies with time and coordinates is derived by the source-function method based on the elementary solution of an instantaneous point contaminant source.

Considering the significant differences in particle size and dielectric property between HMs and SPs, a theoretical model describing their cotransport was developed. The results show that temperature and seepage velocity have a negligible effect on the transport of individual HMs and that the recovery ratio of Cd^{2+} is higher than that of Pb^{2+}, which can be attributed to the lower adsorption of Cd^{2+} on the solid matrix.

1.3.6 Chapter 7: Granular Thermodynamic Substance Flow Model

Based on the energy dissipation caused by the consolidation deformation of porous media under external force and the transport of internal suspended substances, a coupled multi-substance flow (CMF) model was established. This model introduced the new concepts, such as granular temperature and granular entropy, to describe energy dissipation at meso-level. This model used a potential energy density function and migration coefficients to establish the corresponding intrinsic connection between the dissipative force and dissipative flow. This viewpoint unifies the deformation, seepage, and suspended substance transport under the framework of granular thermodynamics. Not only does it reflect the evolution of effective stress in the solid matrix of multicomponents in a particle-reorganized state, but it also allows for the consideration of temperature driving effect.

The proposed CMF model is validated using the experimental results under coupled transport of HMs and SPs. The CMF model is expanded in unsaturated soils driven by coupled seepage–temperature including the transport of HMs. The results clarified the influences of phase transition and attachment–detachment processes resulting from coupled transport, laying the foundation for further implementation in engineering practices.

CHAPTER 2

Experiments on the Transport of Suspended Substances

2.1 MAIN CONTENT OVERVIEW

This chapter experimentally examines the influence of various physical factors, such as particle size, particle concentration, seepage velocity, and other factors, on the transport of suspended particles (SPs, namely, insoluble suspended substances) within porous media through a column seepage test. Traditional theoretical models are applied to simulate the breakthrough curves (BTCs) of SPs, particularly for the case of pulse-injection. From experimental outcomes, variations in key transport parameters are evaluated, such as particle transport velocity, longitudinal dispersivity, detachment coefficient, and recovery rate.

In addition, the temperature and alkalinity on the transport of SPs are also investigated. Changes in transport parameters are analyzed under different seepage velocities and different alkalinity levels. This chapter further explores the influence of test methods, including the injection duration (e.g., instantaneous or continuous patterns) on the seepage transport.

2.2 CLASSICAL TRANSPORT EQUATION

2.2.1 Governing Equations

The transport of suspended substances (e.g., heavy metal ions, viruses, bacteria, silicon powders, red mud particles) in porous media involves complex hydrodynamic processes, including both mechanical dispersion and molecular diffusion. In accordance with the law of mass conservation, the governing equations for convection–diffusion in porous media can be expressed as follows:

$$\frac{\partial(nC)}{\partial t} = \nabla(n \cdot \boldsymbol{D} \cdot \nabla C) - \nabla(n \cdot \boldsymbol{U} \cdot C) \tag{2.1}$$

DOI: 10.1201/9781003783749-2

Under one-dimensional transport conditions, it is assumed that porosity remains constant during the transport processes. Therefore, Eq. (2.1) can be reformulated as follows:

$$\frac{\partial C}{\partial t} = D \cdot \frac{\partial^2 C}{\partial x^2} - u \cdot \frac{\partial C}{\partial x} - \frac{\rho_s}{n} \cdot \frac{\partial \sigma}{\partial t} \tag{2.2}$$

where $\boldsymbol{U}$ is the flow velocity vector in the porous medium; $\boldsymbol{D}$ is the dispersion coefficient tensor; C represents the concentration of suspended substances; n is the porosity of porous medium; t denotes time; x is the transport distance of the suspended substances; $D = \alpha_d \cdot u$ is the hydrodynamic dispersion coefficient, and α_d is the hydrodynamic dispersivity and u is the average internal velocity of the suspended substances; σ stands for the attachment concentration (i.e., adsorption concentration) of the suspended substances, and ρ_s is the solid matrix density.

2.2.2 Classical Isothermal-Adsorption Model

For the classical nonlinear Freundlich isothermal-adsorption, the attachment concentration is expressed by

$$S = K_f C^{\frac{1}{\chi}} \tag{2.3}$$

where S denotes the equilibrium attachment concentration, K_f is the Freundlich adsorption equilibrium constant, and χ is a constant reflecting attachment characteristics.

According to Eq. (2.3), $\partial S / \partial C = (K_f / \chi) \cdot C^{(1-\chi)/\chi}$. In the Freundlich isothermal-adsorption model, the solid matrix is assumed to have an infinite attachment-capacity, which clearly does not align with the facts. To overcome this, the nonlinear Langmuir isothermal-adsorption model incorporates the maximum attachment-capacity of the porous media. It is expressed as

$$S = \frac{K_l C}{1 + K_l C} \cdot S_{max} \tag{2.4}$$

where K_l is the Langmuir constant, with higher values of K_l indicating greater attachment-capacity of the porous media, and S_{max} is the maximum attachment concentration.

From Eq. (2.4), one has $\partial S / \partial C = K_l / (1 + K_l \cdot C)^2 \cdot S_{max}$.

2.3 SEEPAGE TRANSPORT OF SPs WITH DIFFERENT SIZES

2.3.1 Experimental Methods and Materials

2.3.1.1 Experimental Apparatus

The direction of seepage (e.g., horizontal, downward, and upward) influences the seepage transport of SPs. The gravity effect is linked to the particle size and the matrix structure, interacting with other effects, such as the hydrodynamic force, attachment process, and the presence of preferential pathways. The horizontal flow, akin to that observed in an aquifer, is often regarded as preferential. Unfortunately, the deposition process in this case

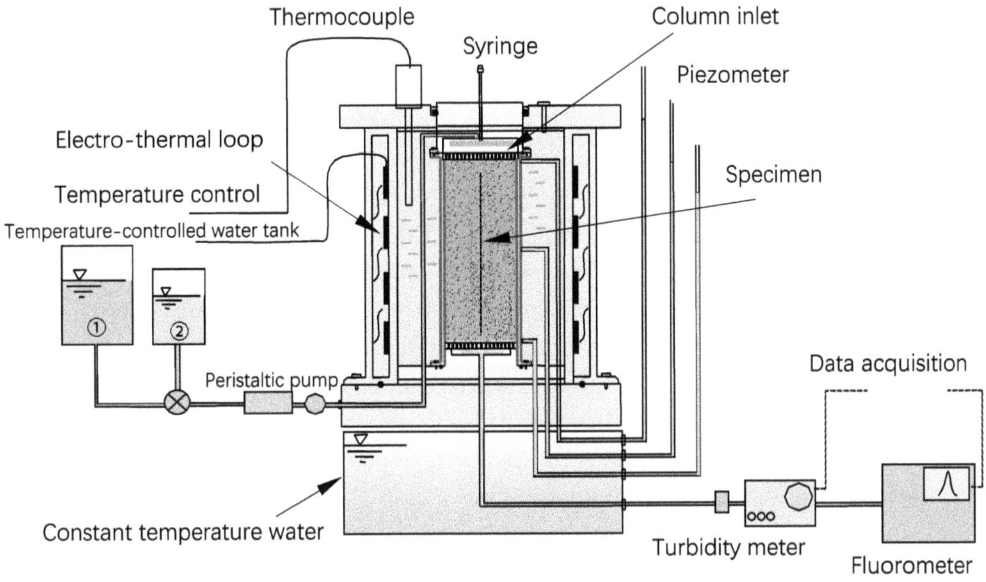

FIGURE 2.1 Experimental setup with temperature control.

is not strictly one dimensional. Upward flow is also a consideration, as packed columns are traditionally saturated with upward flow to reduce air entrapment, although maintaining a one-dimensional uniform particle injection is challenging. It appeared that under down-flow conditions, the transport velocity of SPs was greater than in the horizontal or upward flow scenarios, leading to an underestimation of SPs attachment. However, as the seepage velocity (or the hydrodynamic force) increased, the gravity effect became less significant. Laboratory tests were conducted using a toughened Plexiglas column measuring 300 mm in length and 80 mm in internal diameter (Figure 2.1), which was filled with porous material (i.e., quartz sands). To simplify the setup, water was pumped downward through the vertical column at a consistent velocity (Bai et al., 2016). The column was situated within a temperature-controlled cylindrical chamber made of plexiglass, with temperature regulation achieved by an electrothermal loop inside the chamber and connected to a thermocouple. The test chamber was engineered to withstand temperatures ranging from 10 °C to 80 °C. Nylon membranes were installed at both ends of the sand column to ensure uniform water flow. A temperature-controlled water tank was utilized for water supply, while a peristaltic pump maintained a constant flow rate. Seepage velocity was monitored with a flowmeter positioned at the column inlet. The SPs were injected at the column inlet via a 60-ml syringe. Following each injection, the turbidity of the effluent was measured using a turbidity meter.

2.3.1.2 SPs and Porous Materials

Quartz sands with a particle size in the range of 1–3.5 mm (coefficient of curvature $C_c = 1.03$, uniformity coefficient $C_u = 1.48$) served as porous materials. Prior to the experiments, the sands were treated with 0.1 mol/L H_2SO_4 for 24 h to eliminate impurities, thoroughly rinsed several times with deionized water, and oven-dried at 105 °C for 48 h. To avoid air entrapment, the saturated sand column was packed with quartz sands in 3 cm increments, mixing it with deionized water, and gently vibrated by hand for uniformity. The water level was maintained several centimeters above the sand surface. The solid phase density was $\rho_s = 2.65 \pm 0.05\ \text{g/cm}^3$, with a porosity of $n = 45.1\%$, defined as the ratio of the void volume inside the sand column to the overall column volume.

Nine commercial types of natural spherical SiO_2 particles were selected as the injected SPs, with a particle size distribution (PSD) ranging from 1.02 μm to 47 μm. These particles have a density of 2.26 g/cm^3, and their median diameters (D_{50}) are recorded as $D_{50} = 1.02$, 3.40, 8.31, 10, 15, 20, 25, 33, and 47 μm, respectively. Additionally, a mixture composed of 50 wt% 10-μm particles and 50 wt% 47-μm particles was analyzed, revealing a distinct bimodal PSD with an extended shoulder spanning approximately 4–50 μm, contrasting sharply with the PSD of the nine single particles.

2.3.1.3 Experimental Process

Starting with an initial concentration of 1 mg/ml, the particle mixture underwent dilution through sequential injections of 50 ml of deionized water. Turbidity measurements were taken immediately after each injection. The experiment was deemed complete once the turbidity of the mixture approached zero. Each experiment was conducted a minimum of three times, with the average values used to reduce experimental errors. The nonlinear regression analysis demonstrated a strong correlation between particle concentrations and turbidity levels for the nine particles. The relationship was given by the following equation:

$$C = aN^2 + bN + C \tag{2.5}$$

where C represents SPs concentration, N indicates the turbidity level, and a, b, and c are fitting parameters.

The SPs were injected at the column inlet within a very short time of approximately 2 s (i.e., a "pulse-injection" pattern, or an instantaneous injection pattern). The turbidity in the effluent was measured three times and the averages were computed to minimize errors. The particle concentration was calculated using Eq. (2.5). After each trial, quartz sand was extracted from the column, thoroughly rinsed with deionized water to eliminate impurities, and then repacked into the column. All experiments were conducted under controlled temperatures at 22–24 °C.

2.3.2 Experimental Results

2.3.2.1 Effect of Particle Size

The experiments were conducted for each particle type at various seepage velocities (v = 0.087, 0.173, 0.260 cm/s). In each trial, 30 ml of particle suspension was injected using a syringe, followed by concentration measurements in the effluent at intervals of 3–9 s.

To describe the temporal progression of BTCs, the pore volume (V_P) was introduced (Bai et al., 2017a). This metric represents the ratio of the total volume of water flowing through the porous medium column to the void volume of that column. This dimensionless parameter allows for the presentation of BTCs across different void volumes and seepage velocities on a single figure, facilitating broader conclusions.

The relative concentration, denoted C_R, is defined by the following expression (Bai et al., 2017b):

$$C_R = \frac{C_{out}V_0}{m} \tag{2.6}$$

where C_{out} indicates the particle concentration at the outlet; V_0 represents the void volume of the entire soil column; and $m = C_{inj} \cdot V_{inj}$, with C_{inj} denoting the concentration of injected SPs and V_{inj} being the volume of the particle suspension for each injection.

For a pulse-injection pattern, the concentration $g(t)$ can be expressed in the following forms:

$$g(t) = C_0 \cdot [u(t) - u(t - t_0)] \tag{2.7}$$

$$C_0 = \frac{m}{V_{inj} + Qt_{inj}} \tag{2.8}$$

$$t_0 = \frac{V_{inj} + Qt_{inj}}{Q} \tag{2.9}$$

where C_0 represents the concentration of SPs at the top surface of the column, t_0 denotes the application duration, $u(t)$ is the Heaviside function, $Q = vA$ is the flow rate, and t_{inj} is the duration for which the particle injection occurs.

Equation (2.8) simplifies to $C_0 = C_{inj}$ as $t_{inj} \to 0$ (i.e., pulse-injection), and to $C_0 = m/(Q \cdot t_{inj})$ as $t_{inj} \to \infty$ (constant concentration injection). Figure 2.2 illustrates the experimental BTCs for various particle sizes when $C_{inj} = 0.2$ mg/ml. From Figure 2.2, it is evident that as pore volume (V_P) increases, the relative concentration (C_R) rises until it reaches a maximum value before gradually declining to zero. At a constant seepage velocity, the peak values diminish, while the corresponding V_P shows a slight increase as the SPs size increases. For instance, at a seepage velocity of 0.087 cm/s, the peak values range from $C = 0.029$ mg/ml to 0.005 mg/ml, with corresponding V_P of 0.86–0.94 (Figure 2.2) for $D_{50} = 1.20 - 47\,\mu m$. As particle size increases, the number of pores in the porous medium that are larger than the SPs decreases, while the number of SPs trapped in throat pores increases. This results in a decrease of the peak values in the effluent. Furthermore, the influence of particle size on the transport process diminishes as particle size increases, particularly when comparing $D_{50} = 33\,\mu m$ and $D_{50} = 47\,\mu m$.

2.3.2.2 Effect of Seepage Velocity

When the concentration of the injected SPs is held constant, both the peak values and the corresponding V_P increase as the seepage velocity increases (e.g., $C_{inj} = 0.2$ mg/ml; refer to Figure 2.2). For instance, at $D_{50} = 10\,\mu m$, peak concentrations are recorded as

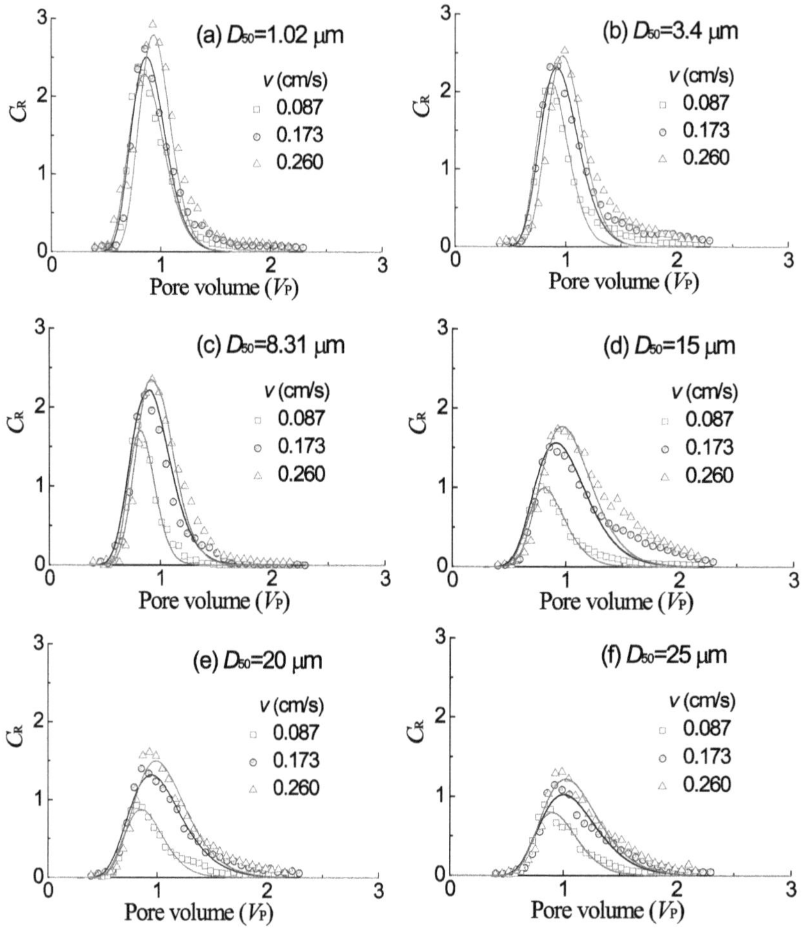

FIGURE 2.2 BTCs for different particle sizes. (a) 1.02 μm. (b) 3.4 μm. (c) 8.31 μm. (d) 15 μm. (e) 20 μm. (f) 25 μm.

$C = 0.015, 0.024$, and 0.028 mg/ml, with associated $V_P = 0.92$, 0.98, and 1.06 (Figure 2.3) for v = 0.087, 0.173, and 0.260 cm/s, respectively. In fact, the increase in water seepage velocity raises the hydrodynamic force acting on SPs, leading to a reduction in both deposition and gravitational effects. Consequently, there is an increase in the number of SPs at the outlet, reflected in higher peak values. However, compared to the seepage velocity of the water, the seepage velocity of SPs is noticeably lower due to the intense collision between the SPs and the porous medium matrix. Higher seepage velocities result in a greater likelihood of the collision. Consequently, the transport paths of SPs become extended due to the detours encountered within the porous media, resulting in a noticeable delay in the V_P corresponding to the peak value of BTCs.

2.3.2.3 Effect of Injection Concentration

Figure 2.3 presents the BTCs corresponding to various injection concentrations ($C_{inj} = 0.2 - 4.0$ mg/ml) and seepage velocities (v = 0.087, 0.173, and 0.260 cm/s), with a

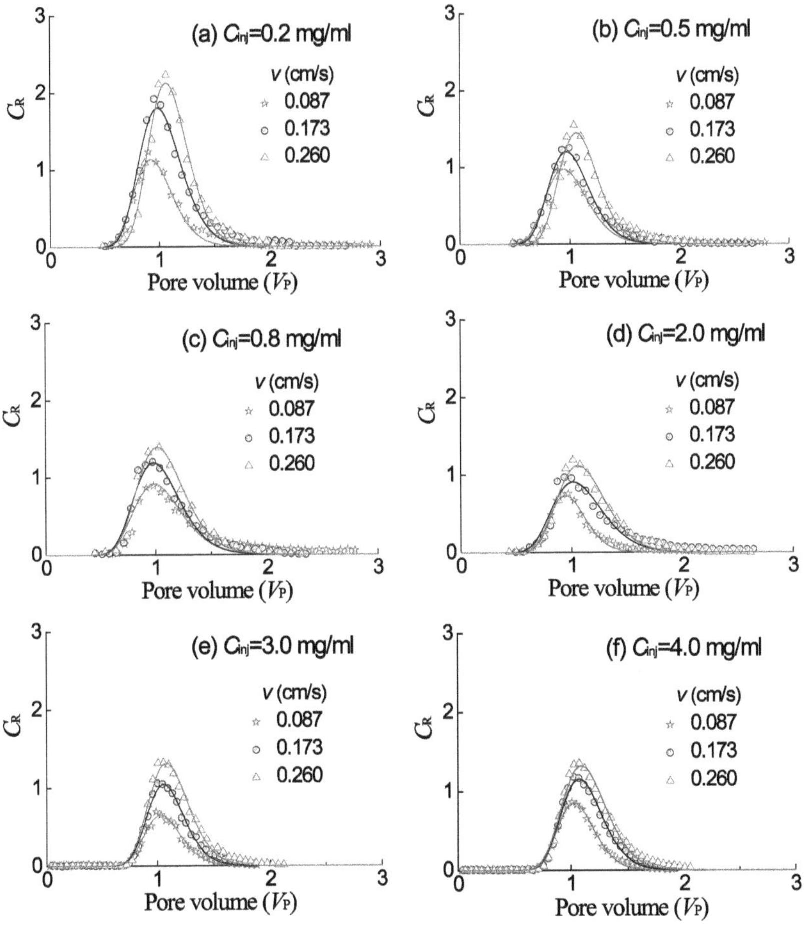

FIGURE 2.3 BTCs under varying injection concentrations. (a) 0.2 mg/ml. (b) 0.5 mg/ml. (c) 0.8 mg/ml. (d) 2.0 mg/ml. (e) 3.0 mg/ml. (f) 4.0 mg/ml.

particle size of $D_{50} = 10\ \mu m$. Observations indicate that, for a fixed seepage velocity, the peak values of BTCs rise as the concentration (C_{inj}) of the injected SPs increases; concurrently, the respective V_P exhibit a slight increase. For instance, at $v = 0.087$ cm/s, the values are $C = 0.015$, 0.032, 0.044, and 0.092 mg/ml, leading to $V_P = 0.9 - 1.01$ for $C_{inj} = 0.2$, 0.5, 0.8, and 2.0 mg/ml, respectively. However, the relative concentration diminishes ($C_R = 1.23$, 1.05, 0.89, and 0.76 for $C_{inj} = 0.2$, 0.5, 0.8, and 2.0 mg/ml, respectively). Similar patterns are evident for $v = 0.173$ and 0.260 cm/s. As the concentration of the injected SPs increases, there is a corresponding rise in both the number of SPs that are deposited onto the porous medium matrix and the frequency of collision between the SPs and the matrix surface. This phenomenon causes a reduction in the permeability of the porous medium due to void plugging from the SPs deposited onto the porous medium matrix.

2.3.3 Discussions on Transport Parameters

2.3.3.1 Influence of Particle Size

By utilizing the optimization analysis, this model enables the determination of four fitting parameters (u, α_d, k_d, and k_r). The α_d is expressed as follows:

$$\alpha_d = \frac{D}{u_0} \tag{2.10}$$

where $u_0 = v/n$, and u_0 is the average interstitial seepage velocity.

However, fitting experimental data to a nonlinear model poses considerable difficulties. When more than three parameters need to be estimated, achieving unique and simultaneous estimates for all parameters becomes highly challenging. However, since the detachment coefficient k_r (i.e., release coefficient) is considerably smaller than the attachment coefficient k_d (i.e., deposition coefficient) and is predominantly associated with the later stage of the transport process, it can be assumed that $k_r = 0.1k_d$. This simplification reduces the number of parameters requiring identification to three.

Referring to the simulated BTCs in Figure 2.2 ($R^2 > 0.91$), Figure 2.4a illustrates how the ratio of the interstitial particle velocity to that of interstitial water changes ($u_r = u/u_0$) with the particle size across three different seepage velocities. It is evident that for the transport of small SPs (e.g., $D_{50} = 10$ μm), the transport velocity exceeds that of the water velocity (i.e., $u_r > 1$), particularly in scenarios involving low seepage velocities (e.g., v = 0.087 cm/s). This behavior can be explained by the size exclusion effect. However, as the particle size increases, their transport velocity becomes less than the water velocity (i.e., $u_r < 1$), primarily due to the logging effect. The transport process of SPs is mainly influenced not only by the hydrodynamic process, in addition to SPs size, but also by other factors such as the pore structure of the matrix and more. For smaller SPs, the transport velocity exhibits a clear size exclusion effect, as the SPs preferentially follow central streamlines. In contrast, the larger SPs tend to move more irregularly due to pore size constraints and the increased number of SPs being trapped in throat pores, which consequently results in a decrease in transport velocity. Moreover, higher seepage velocity exacerbates the irregular movement of SPs, further diminishing their transport velocity. These seemingly contradictory outcomes are a result of complex coupling effects involving the pore structure of the matrix, gravity effect, size exclusion effect, hydrodynamic dispersion, SPs concentration, and the electrostatic force.

Figure 2.4b illustrates that dispersivity is strongly dependent on SPs size, highlighting a positive relationship between dispersivity and the median particle diameter. This indicates that the extent of dispersion at any seepage velocity is influenced by both pore space geometry and colloid size. In fact, the transport of SPs is mainly governed by hydrodynamic processes and the pore structure of the matrix. Larger SPs tend to move more irregularly, constrained by pore size limitations, resulting in greater dispersivity. Small SPs (e.g., $D_{50} = 10$ μm) show a decreasing dispersivity as velocity increases (e.g., v = 0.087 to 0.173 then to 0.260 cm/s) due to their preferential advection along the central streamlines. However, at higher seepage velocities (e.g., v = 0.260 cm/s), the hydrodynamic dispersivity becomes more dominant as particle size increases, as shown in Figure 2.4b, due to enhanced collision between the matrix surface and the SPs.

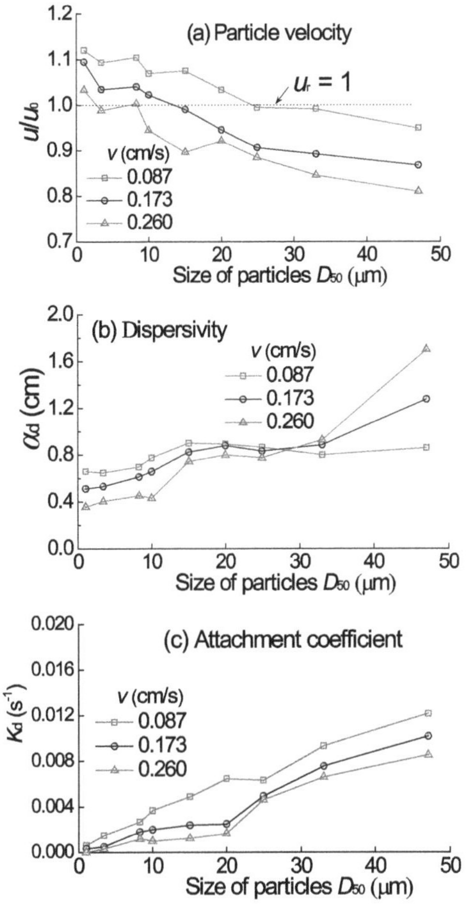

FIGURE 2.4 Transport parameters with particle size. (a) Particle velocity. (b) Dispersivity. (c) Attachment coefficient.

From Figure 2.4c, it is evident that larger SPs exhibit higher attachment coefficients than smaller SPs. This can be attributed to the larger SPs being more restricted by the narrow pores of the porous medium. Additionally, Figure 2.4c demonstrates that as seepage velocity increases, the attachment coefficient decreases, and less deposition occurs for both higher seepage velocities and smaller SPs (e.g., v = 0.260 cm/s and $D_{50} < 10$ μm).

2.3.3.2 *Influence of SPs Concentration*

Figure 2.5a illustrates that the SPs with a unimodal PSD ($D_{50} = 10$ μm) display a clear "preferential path effect" under injection concentration. This effect is particularly pronounced at lower seepage velocities (e.g., v = 0.087 cm/s). As the injection concentration rises, the transport velocity of these particles decreases and gradually stabilizes. In contrast, for mixed SPs with a bimodal PSD, the transport velocity initially increases before gradually decreasing to a steady state as the injection concentration continues to rise. This behavior results from the combined influence of the "preferential path effect" acting on small SPs and the "detours path effect" associated with larger SPs. Figure 2.5b further

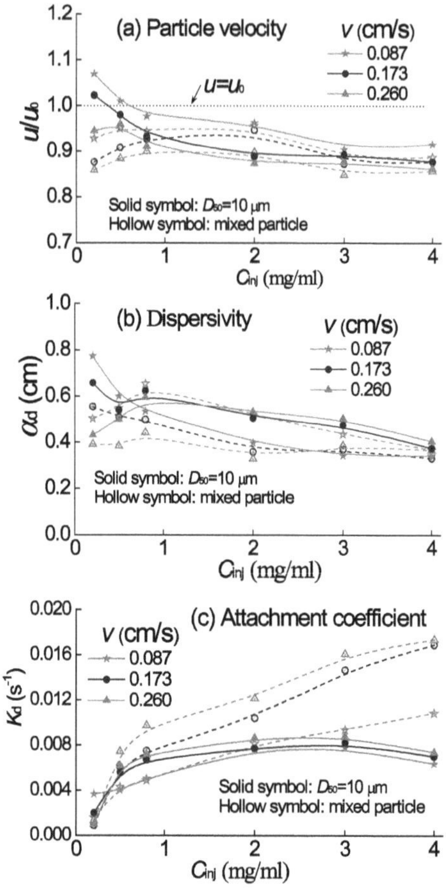

FIGURE 2.5 Transport parameters with the injected concentration. (a) Particle velocity. (b) Dispersivity. (c) Attachment coefficient.

shows that although the relationship between the variation of the dispersivity and the concentration of injected SPs is intricate, it generally trends downward toward a steady state as the injection concentration increases.

Figure 2.5c demonstrates that the attachment coefficient for mixed SPs rises steadily as the injection concentration. Notably, the mixed SPs with a larger median diameter demonstrate higher attachment coefficients compared to those with a smaller median diameter (e.g., $D_{50} = 10\ \mu\text{m}$). For SPs with $D_{50} = 10\ \mu\text{m}$, the attachment coefficients first increase but can stabilize or even slightly decrease as injection concentration rises. Consequently, the SPs captured by the matrix may diminish or stop once the drag force exerted by the moving water surpasses the attractive normal force acting on the SPs.

2.4 EFFECT OF TEMPERATURE ON SEEPAGE TRANSPORT

2.4.1 Experimental Description

The experiments utilized quartz sands with a median diameter of $d_g = 1.18$ mm and a size range of 0.5–2 mm as the porous material. The proportion of quartz sand particle in the size ranges of 0.5–1 mm and 1–2 mm were approximately 41.5% and 58.5%, respectively. The density of the solid matrix was $\rho_s = 2.65 \pm 0.05\,\text{g/cm}^3$, and the porosity was n = 42.5%. A total of six types of commercial natural SiO_2 particles were injected, featuring median diameters (D_{50}) of 10, 15, 20, 25, 33, and 47 μm and a density of $\rho_s = 2.26\,\text{g/cm}^3$, respectively. The scanning electron microscope (SEM) images of two representative SPs ($D_{50} = 10$ and $25\,\mu\text{m}$) are shown in Figure 2.6.

SPs were injected at the column inlet in a rapid pulse-injection pattern lasting about 2 s (Bai et al., 2017a). To monitor the concentration of a dissolved tracer (DT, that is, fluorescein), a fluorometer was utilized during typical tests. The instrument can detect fluorescein concentration as low as 0.1 ppb (parts per billion, that is, 10^{-4} g/L). Following each experiment, the quartz sand was extracted from the column, washed multiple times with deionized water to eliminate impurities, and then repacked into the column.

2.4.2 Results and Discussion

2.4.2.1 Effect of Temperature

The experimental BTCs for various particle sizes ($D_{50}/d_g = 0.0085 - 0.0398$) are presented in Figure 2.7. It is clear that the peak values of BTCs decline as particle size increases, while they rise with higher seepage velocity. Additionally, as the temperature increases, the peak values are reduced and the corresponding V_P is delayed. The increased dispersivity associated with larger SPs leads to greater spreading, which results in flatter peak values, particularly at higher temperatures. At $D_{50} = 10\ \mu\text{m}$ and v = 0.066 cm/s (Figure 2.7a), the peak values are $C_{out} = 0.129$, 0.105, and 0.064 mg/ml while the corresponding $V_P = 0.95$, 1.0, and 1.05 for T = 15 °C, 35 °C, and 55 °C, respectively.

An increase in temperature intensifies the flow turbulence within the inhomogeneous flow field of the porous medium's pores. Additionally, the thermodynamic driving force rises with temperature due to the stochastic Brownian motion behavior of SPs, making straining deposition more significant. As the temperature rises, the probability of collision increases, and the transport pathways for SPs become longer because of detours within the porous medium. Consequently, the V_P corresponding to the peak value of BTCs is significantly delayed, and the quantity of SPs retained by the throat pore rises, leading to a reduction in the peak values in the effluent.

2.4.2.2 Comparison with Conservative Tracer

The dispersivity of SPs is greater than that of the DT (Figure 2.8), as reflected in the shapes of BTCs at different temperatures ($D_{50} = 25\ \mu\text{m}$). The rapid transport of SPs is primarily driven by the size exclusion effect, which directs SPs to be transported through preferential pathways. Larger SPs move at slower velocities. Consequently, smaller SPs reach the outlet of the column first. The rapid transport of SPs can also be linked to the velocity distribution within a pore, which can be understood through hydrodynamic chromatography in a cylindrical tube. In a laminar flow regime, the velocity profile at a microscopic level exhibits a parabolic shape, with the highest velocity occurring at the tube's axis. Actually, large SPs are only transported near the axis of the tube at higher velocities, whereas the DT tends to move closer to the wall at a lower water velocity.

FIGURE 2.6 SEM micrographs showing injected SPs. (a) $D_{50} = 10\,\mu\text{m}$. (b) $D_{50} = 25\,\mu\text{m}$.

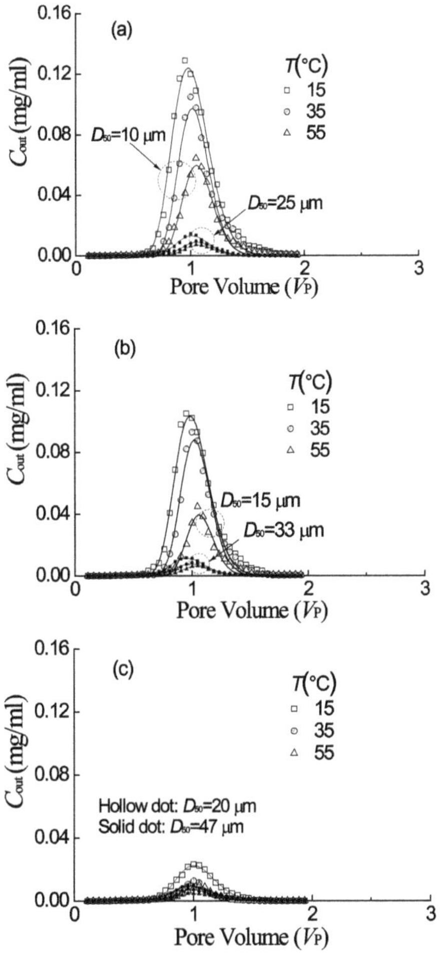

FIGURE 2.7 Particle size effect on BTCs. (a) 10 and 25 μm. (b) 15 and 33 μm. (c) 20 and 47 μm.

At elevated temperatures, the irregular motion of dissolved solutes (i.e., DT) in the moving water within the porous medium is intensified by Brownian motion. This results in a delay in the transport velocity of DT as temperature increases (see Figure 2.8). This can be linked to a rise in the number of transport paths due to the enhanced thermodynamic process. Furthermore, the size exclusion effect that influences the transport of larger SPs in the moving water is diminished at higher temperatures. Consequently, at higher temperatures, both DT and SPs experience longer transport paths due to the detours within the porous medium before arriving at the outlet, and the transport velocity of DT remains consistently lower than that of SPs throughout the investigated temperature range.

2.4.3 Discussion of Transport Parameters

Figure 2.9a presents the relationship between the interstitial particle velocity and interstitial water velocity (u/u_0) against temperature for various SPs (D_{50} = 25 and 47 μm) and seepage velocities (v = 0.066 and 0.199 cm/s). It indicates that the transport velocity of SPs

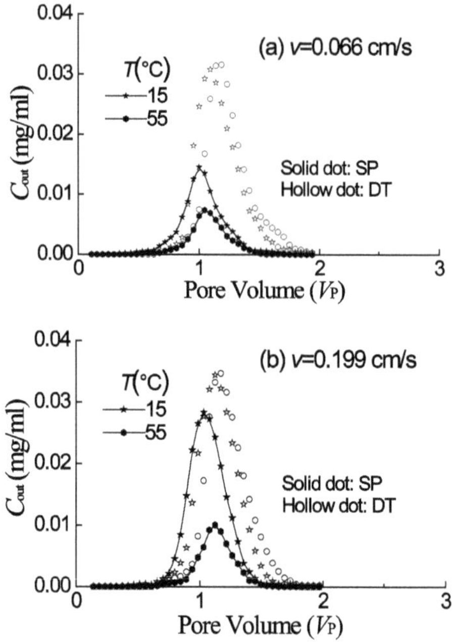

FIGURE 2.8 BTCs contrasting conservative tracer with SPs. (a) $v = 0.066$ cm/s. (b) $v = 0.199$ cm/s.

is lower than the water velocity ($u/u_0 = 0.85 - 0.9 < 1$), particularly for the $D_{50} = 47$ μm SPs, which is attributed to the logging effect. This differs from the transport process of smaller SPs ($u/u_0 > 1$), where the size exclusion effect plays a more dominant role than the detours in the transport paths. From Figure 2.9b, the transport velocity of SPs decreases as the particle size increases, across varying seepage velocities (v = 0.066 and 0.199 cm/s) and temperatures (T = 15 °C and 55 °C).

An increase in temperature leads to stronger repulsive interaction due to the rise in total interaction energy, such as repulsive electrical double-layer forces, as described by the DLVO theory. This enhancement increases collisions between SPs and between SPs and the surface of the porous matrix, ultimately reducing the size exclusion effect induced by preferential streams. Additionally, higher temperatures accelerate the irregular movement of SPs through the Brownian motion, which extends the transport paths and thereby decreases the transport of SPs (Figure 2.9a). Within the examined temperature range of 15–55 °C, a negative correlation exists between u/u_0 and T, although the evolution process varies somewhat at low water velocities (e.g., v = 0.066 cm/s). Generally, larger SPs experience more irregular movement due to the limitations imposed by pore size, particularly at higher temperatures. Figure 2.9a also demonstrates that the transport velocities of SPs consistently exceed those of DT across different seepage velocities and temperatures, further highlighting the influence of preferential streams.

The dispersion coefficient (D) can be expressed by the sum of two components: $D = D_m + D_e$, where D_m is mechanical dispersion and D_e is molecular diffusion:

$$D = D_m + D_e = \alpha_d u_0^{\lambda} + \frac{D_{AB}}{\tau_0} \tag{2.11}$$

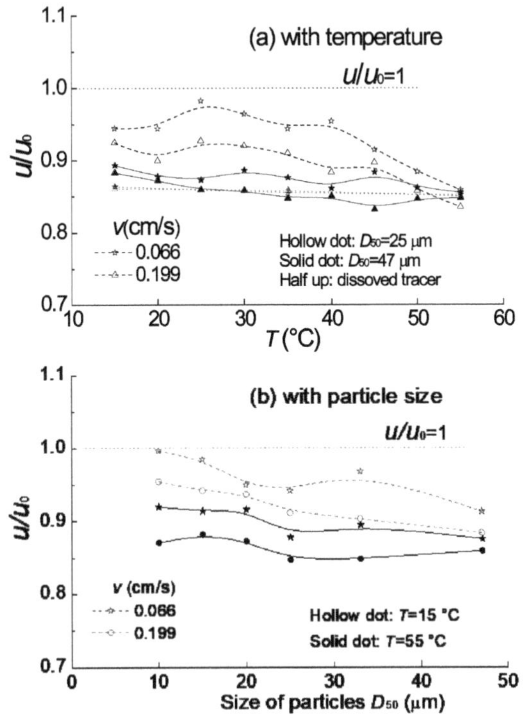

FIGURE 2.9 SPs velocity against (a) temperature and (b) particle size.

where D_{AB} denotes the molecular diffusion coefficient, the exponent λ is an empirical coefficient, and τ_0 indicates the tortuosity of the porous medium.

For spherical particles, the molecular diffusion coefficient can be calculated using the Stokes–Einstein diffusion equation:

$$D_{AB} = \frac{k_B T_{abs}}{3\pi\mu_B d_p} \tag{2.12}$$

where d_p represents the particle diameter; $T_{abs} = T + 273$ is the absolute temperature (K), with T being the temperature in Celsius (°C); μ_B is the water dynamic viscosity at 25 °C (i.e., 0.89×10^{-3} (N·s)/m^2); and k_B is the Boltzmann constant (1.38×10^{-23} J/K).

Equation (2.12) clearly illustrates that the molecular diffusion coefficient decreases sharply as particle size increases, indicating that larger SPs exhibit a smaller molecular diffusion coefficient. Although temperature positively influences the molecular diffusion coefficient, its effect is somewhat restricted due to the narrow temperature range (15–55 °C). When the Peclet number (p_e) is significantly large for the DT, such as fluorescein ($p_e = 1.8 \times 10^3 - 5.5 \times 10^3$), mechanical dispersion becomes the dominant factor. Consequently, the molecular diffusion (D_{AB}) of SPs can be considered negligible, leading to the conclusion that the hydrodynamic dispersion coefficient of SPs is mainly governed by its mechanical dispersion. In the mechanical dispersion term outlined in Eq. (2.11), the empirical coefficient λ typically ranges from 1 to 1.3 and is commonly taken as $\lambda = 1$. Therefore, the hydrodynamic dispersion coefficient is given by $D = \alpha_d \cdot u_0$, with the longitudinal dispersivity α_d ($\alpha_d = D/u_0$) regarded as a length dimension.

Figure 2.10 demonstrates a close relationship between longitudinal dispersivity α_d, temperature and particle size. From Figure 2.10a, the longitudinal dispersivity of SPs (e.g., $D_{50} = 25$ and $47\ \mu m$) initially decreases slightly with an increase in temperature up to T = 30 °C, which corresponds to a reduction in the transport velocity of SPs, and then remains nearly constant with a further temperature increase. In practice, the primary determinant of the hydrodynamic dispersion coefficient (D) is the average interstitial particle velocity (u), as opposed to the average interstitial fluid velocity (u_0). Consequently, using the simplified equation $\alpha_d = D/u_0$ likely underestimates dispersivity α_d, particularly as temperature rises, since u_0 represents the superficial velocity ($u < u_0$ for SPs). Additionally, the observed decrease in dispersivity with increasing temperature is associated with the deposition of larger SPs, which indicates an increase in the attachment coefficient, as depicted in Figure 2.11a.

The dispersivity of SPs, which α_d ranges from 0.19 cm to 0.45 cm, is typically greater than that of DT, which α_d ranges from 0.19 cm to 0.26 cm, for SPs with sizes $D_{50} = 25$ and $47\ \mu m$, particularly at lower temperatures (between T = 15 °C and 30 °C; Figure 2.10a). However,

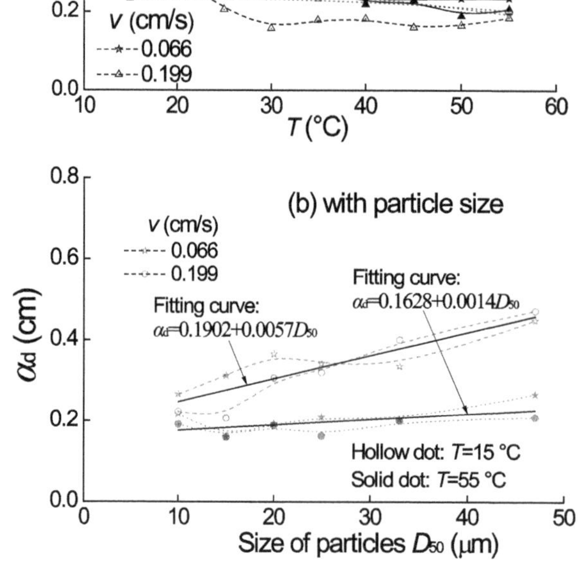

FIGURE 2.10 Dispersivity against (a) temperature and (b) particle size.

this difference diminishes as temperature increases, suggesting that the dispersion coefficients of DT differ fundamentally from those of SPs.

Figure 2.10b illustrates that the influence of seepage velocity on the dispersivity is unclear across a broad diameter range of 10–47 μm. To better understand the relationship between particle size and longitudinal dispersivity, the regression curves were drawn for each temperature (i.e., T = 15 °C and 55 °C), independent of seepage velocity. A positive correlation exists between hydrodynamic dispersivity α_d (cm) and medium diameter D_{50} (μm):

$$\alpha_d = d_1 + d_2 D_{50} \tag{2.13}$$

where d_1 and d_2 represent the fitting parameters.

The values for d_1 and d_2 are 0.1902 and 0.0057 ($R^2 = 0.81$) at T = 15 °C, and 0.1628 and 0.0014 ($R^2 = 0.34$) at T = 55 °C, respectively. The rate of increase in longitudinal dispersivity with particle size is much slower at high temperatures compared to low temperatures.

Generally, the attachment coefficient tends to increase with flowing velocity up to a critical threshold, after which it begins to decline. This influence on the attachment coefficient is complex, as it is affected by factors such as the properties of SPs, alkalinity, the matrix structure, temperature, and more. As temperature increases, larger aggregates may form, potentially bridging and plugging the pore throats due to particle mobilization. This results in longer transport paths for particles within the porous medium, further delaying the penetration of SPs. Consequently, the number of the captured SPs increases, ultimately elevating the attachment coefficient, particularly at higher seepage velocity (e.g., v = 0.199 cm/s; see Figure 2.11a).

Figure 2.11b clearly illustrates that the attachment coefficients for larger SPs are higher than those for smaller SPs. The attachment coefficient approaches nearly zero when the median diameter (D_{50}) is below a certain threshold (e.g., $D_{50} = 10$ μm at v = 0.066 cm/s and T = 15 °C). This threshold diminishes with increases in temperature and seepage velocity. It is quantified by the ratio D_{50}/d_g, which is 0.008 when T = 15 °C. Below this threshold, the particle size is negligible compared to the pore throat, resulting in fewer SPs being able to deposit in narrow or disconnected pores, while the majority can migrate freely with the water.

Figure 2.11b indicates that the attachment coefficient increases gradually when the median diameter exceeds another critical value ($D_{50} = 25$μm; that is, corresponding to $D_{50}/d_g = 0.021$), beyond which it remains relatively constant. This reveals a distinct demarcation in the curve, attributable to the limitations imposed on larger SPs by the narrow pores. At this time, fewer SPs can migrate long distances due to particle clogging within the porous medium, leaving the regions far away from the inlet largely unobstructed. Notably, this critical value varies depending on factors such as temperature, concentration, flow rate, particle type, and porous media. Additionally, the relationship between k_d and D_{50} can thus be expressed in the form of the Boltzmann function, as shown in Figure 2.11b:

$$k_d = k_{d\max}\left[1 - \frac{1}{1 + \exp^{(D_{50}/d_g - x_0)/\beta}}\right] \tag{2.14}$$

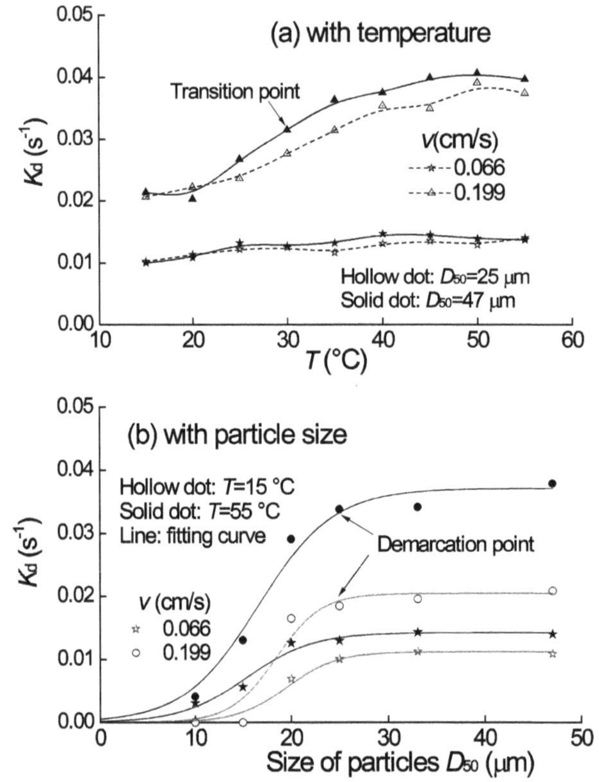

FIGURE 2.11 Attachment coefficient with (a) temperature and (b) particle size.

where $k_{\mathrm{d\,max}}$ is the maximum value of k_{d}, x_0 is the center value, and β is the shape parameter.

Equation (2.14) effectively describes the attachment coefficient across a broad spectrum of particle sizes. According to this equation, k_{d} is zero when $D_{50} = 0$, and reaches its maximum value, $k_{\mathrm{d\,max}}$, when the particle size becomes sufficiently large relative to the pore size $(D_{50} \to d_{\mathrm{g}})$.

Pulse-injection experiments play a key role in assessing the transport of SPs and conveniently the transport parameters. Nevertheless, it is also common in practical applications to carry out long-term injection with concentrations that vary over time.

2.5 EFFECT OF ALKALINITY ON SEEPAGE TRANSPORT

2.5.1 Materials and Methods

Quartz sands with a median diameter *of* $d_{\mathrm{g}} = 1.18$ mm (size range: 0.5–2 mm; containing 41.5% particles of 0.5–1 mm and 58.5% particles of 1–2 mm) were utilized as porous materials. The sand density was $\rho_{\mathrm{s}} = 2.65 \pm 0.05\,\mathrm{g/cm^3}$, and the porosity of the sand column was n = 42.5%, which was carried out layer by layer based on the sand weight and specific density. Transport experiments were carried out using silicon powders in various alkaline suspensions (Bai et al., 2020a).

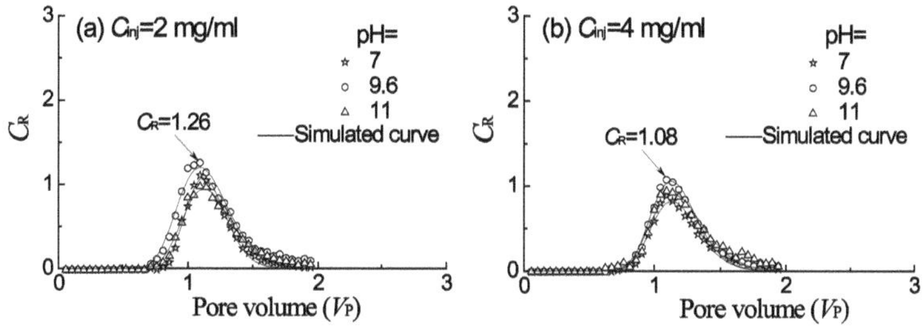

FIGURE 2.12 BTCs of SPs under various alkalinities (Case 1). (a) $C_{inj} = 2$ mg/ml. (*b*) $C_{inj} = 4$ mg/ml

The transport of SPs (i.e., a kind of commercial, natural silicon powders, mainly composed of SiO_2) are investigated at different alkalinity. The pulse-injection pattern was adopted (i.e., a period of 2 s), including two injection cases, which are commonly encountered in practical engineering. The first case involved modulating both the alkalinity of the injected suspension and the seepage water to specific alkalinity levels (pH = 7, 8.2, 9.6, 11; Case 1) by adding 1 mol/L sodium hydroxide (NaOH) solution; while the second case involved adjusting the alkalinity of the injected suspension to different levels (pH = 7, 8.2, 9.6, 11, 13; Case 2), while the alkalinity of the seepage water provided by the water tank was maintained at a neutral with pH = 7.

2.5.2 Experimental Results

For Case 1, the peak values of BTCs generally decline as the injection particle concentration increases (e.g., $C_R = 1.26 \rightarrow 1.08$ when $C_{inj} = 2$ mg/ml $\rightarrow$ 4 mg/ml at v = 0.066 cm/s and pH = 9.6; refer to Figure 2.12). Conversely, the peak values rise with increasing seepage velocity. Initially, the peak values rise as alkalinity increases from pH = 7 to pH = 9.6, and then decline as alkalinity further increases from pH = 9.6 to pH = 11. The corresponding V_P remains nearly the same, ranging from $V_P = 1.06$ to 1.09.

An increase in alkalinity from pH = 7 to pH = 9.6 enhances the repulsive interactions between SPs and between SPs and the matrix, thereby increasing the likelihood of a higher peak value. For instance, at v = 0.066 cm/s (Figure 2.12a), the peak values are $C_R = 1.10$ and 1.26 for pH = 7 and pH = 9.6, respectively. In scenarios with lower alkalinity, the repulsive force rises with increasing alkalinity, even as transport paths lengthen, which diminishes the attachment likelihood of SPs. However, when alkalinity is increased further from pH = 9.6 to pH = 11, the transport paths for SPs become longer due to the collision between SPs and between SPs and the matrix, likely leading to more pronounced deposition. The chemical reaction between the matrix (or SPs) and sodium hydroxide complicates the transport process of SPs. Consequently, more SPs are captured by throat pores, leading to a reduction in the peak values.

For Case 2, the peak values of SPs initially increase and then decrease (C_R = 1.10 $\rightarrow$ 1.17 $\rightarrow$ 1.32 $\rightarrow$ 1.02 for v = 0.066 cm/s and C_{inj} = 2 mg/ml; see Figure 2.13a) with rising alkalinity (pH = 7 $\rightarrow$ 9.6 $\rightarrow$ 11 $\rightarrow$ 13), while the corresponding V_P remain nearly constant ($V_P = 1.04 - 1.08$). This trend is also attributed to the repulsive interactions between SPs

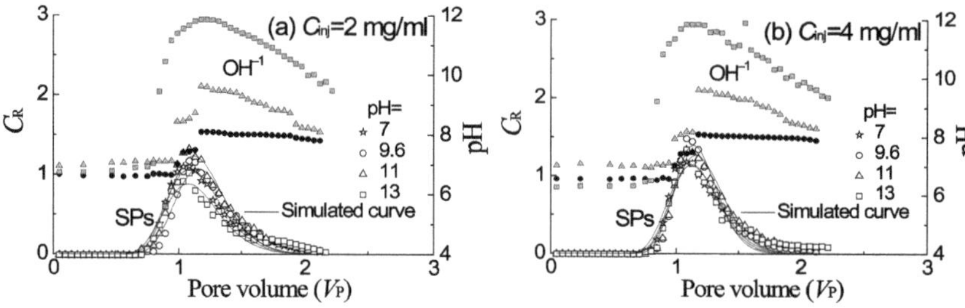

FIGURE 2.13 BTCs of SPs under various alkalinities (Case 2). (a) $C_{\text{inj}} = 2$ mg/ml. (b) $C_{\text{inj}} = 4$ mg/ml.

and between SPs and the matrix. However, in contrast to Case 1, the peak concentrations continuously rise with increasing alkalinity up to about pH = 11 (it is pH = 9.6 for Case 1). For Case 2, the alkalinity of pore water in the porous medium is expected to be lower than that in Case 1 because of the pulse-injection (2 s) of OH^- ions. Therefore, the alkalinity of the pore water during the transport processes of SPs varies continuously (see Figure 2.13). The transport mechanisms of SPs in both cases are fundamentally equivalent.

Figure 2.13 illustrates the transport processes of OH^- ions for various injected concentrations (indicated by pH; e.g., pH = 9.6). These processes are somewhat more intricate compared to those of SPs due to their coupling associated with the OH^- ions. On one hand, OH^- ions that are dissolved in the water travel with the water. Conversely, OH^- ions are released continuously from SPs that move and deposit, as a result of the further cleanout of seepage flow. It is evident that the peak values of OH^- ions at different SPs concentrations (C_{inj} = 2, 4, 6, 8 mg/ml) and seepage velocities (v = 0.066 cm/s) gradually stabilize as the alkalinity increases ($\text{pH} = 7 \rightarrow 9.6 \rightarrow 11 \rightarrow 13$). The corresponding V_P exhibit similar patterns.

Figure 2.13 demonstrates that SPs are transported earlier than OH^- ions across the alkalinity range. This observation is particularly significant when the alkalinity is represented by the concentration of OH^- ion rather than pH value. This finding aligns with numerous prior studies involving SPs in DT despite the strong physicochemical action between OH^- ions and SiO_2 particles as well as the matrix. Due to the preferential advection of SPs along the central streamline, the size exclusion effect of SPs results in a lower dispersion coefficient for the particle plume compared to OH^- ions. In fact, large SPs require high velocities for movement through the middle of the pores, while smaller dissolved OH^- ions can migrate closer to the solid matrix surfaces where the water velocity is low.

2.5.3 Effect of Alkalinity on the Transport Process

Figure 2.14 illustrates how particle velocity (u/u_0), attachment coefficient (k_d), dispersivity (α_d), and recovery rate (R_e) vary with alkalinity (pH) across different seepage velocities (v = 0.066, 0.133, and 0.199 cm/s) in B-spline curves. Here, the recovery rate (R_e) is defined as the ratio of suspended substances in the effluent to the injected suspended substances. Figure 2.14a is evident for SPs with a median diameter of $D_{50} = 10$ μm, the velocity is consistently smaller than the water velocity ($u/u_0 = 0.82 - 0.93 < 1$) due to the clogging effect.

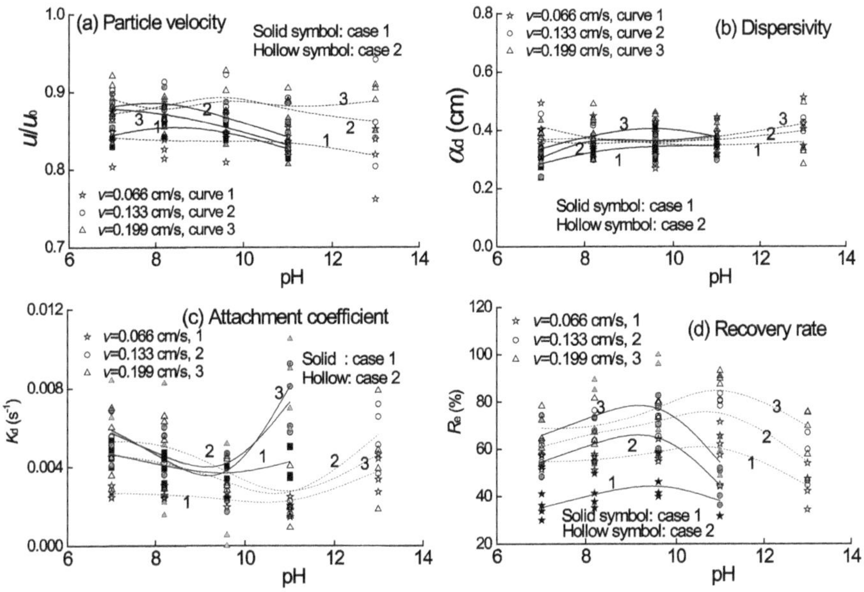

FIGURE 2.14 Transport parameters of SPs as a function of alkalinity. (a) Particle velocity. (b) Dispersivity. (c) Attachment coefficient. (d) Recovery rate.

This behavior contrasts with the transport process of smaller SPs ($u/u_0 > 1$), suggesting that the size exclusion effect plays a more significant role than the detours in the transport pathway. With increasing pH value, there is a slight decrease in particle velocity, which can be explained by the reduction in effective porosity caused by pore plugging, along with an increase in dispersivity (Figure 2.14b).

The dispersivity values exhibit a complex relationship with alkalinity (Bai et al., 2017b, 2020c). However, at a constant seepage velocity, these values remain within a narrow range within the investigated pH range (pH = 7–13). As alkalinity increases, the repulsive interactions between SPs and between SPs and the matrix intensify, which is attributed to the increase in total interaction energy (i.e., the repulsive electrical double-layer forces) by the DLVO theory. Consequently, the coupled effect of reduced particle velocity and increased dispersivity initially leads to a decrease in the attachment coefficient as alkalinity rises from pH = 7 to pH = 11 (Case 2; Figure 2.14c), followed by an increase as alkalinity continues to rise from pH = 11 to pH = 13. In fact, the increase in dispersivity is beneficial at low alkalinity, while the decrease in particle velocity becomes significant at high alkalinity. In general, the variations in attachment coefficient and recovery rate due to alkalinity differences become more pronounced as the seepage velocity increases ($v = 0.066 \rightarrow 0.133 \rightarrow 0.199$ cm/s; Figure 2.14d).

The recovery rate rises with increasing alkalinity until a critical pH is attained, after which it begins to decline. The case of seepage water with varying alkalinity (Case 1; solid symbol in Figure 2.14) is analyzed as a representative example, with the critical pH found to be approximately $(\text{pH})_{\text{cri}} = 9.6$. Certainly, this critical pH is significantly influenced by the properties of SPs and the matrix, such as the attachment-capacity, specific surface area, and structural characteristics. In conditions of low alkalinity ($pH = 7 \rightarrow 8.2 \rightarrow 9.6$, as NaOH concentration increases), a chemical reaction occurs between the SiO_2-based solid matrix (or SPs) and sodium hydroxide (NaOH):

$$SiO_2 + 2NaOH = Na_2SiO_3 + H_2O \tag{2.15}$$

The SiO_3^{2-} ions will strongly adhere to the matrix (or SPs) as a part of the adsorption layer, while Na^+ ions (as counterions) diminish with increasing distance from the matrix (or SPs) surface, gradually dissolving in the pore water (Figure 2.15a: Stage 1). Under alkalinity conditions, the zeta potentials of materials like silica powders, quartz sands, glass beads, and kaolinite particles generally rise as pH increases. The potential energy consists of both repulsion and attraction components. As the thickness of the adsorption layer increases, the repulsive force between the quartz matrix and SPs also rises. Consequently, the dispersivity of SPs increases (Figure 2.14b), while the attachment coefficient decreases (Figure 2.14c), leading to an increase in the recovery rate of SPs in the effluent (Figure 2.14d).

However, as the alkalinity value continues to increase (e.g., $pH = 9.6 \rightarrow 11$ for Case 1), the chemical reaction between SiO_2 and NaOH reaches equilibrium due to the restricted specific surface area of both the matrix and SPs. At this point, the presence of a significant concentration of Na^+ ions in solution (Figure 2.15a: Stage 2) diminishes the surface electric

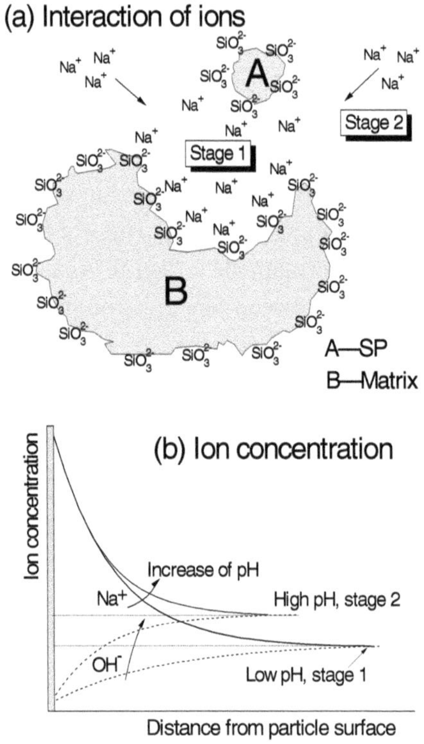

FIGURE 2.15 Electrical double layer resulting from NaOH solution. (a) Interaction of ions. (b) Distribution of ion concentration.

potential between the pore water and the solid particle surface. This results in a decreased thickness of the diffusion layer on the surface of the matrix and SPs (Figure 2.15b: Stage 2). Based on the Gouy–Chapman theory, a relationship exists between the surface charge of the particles σ_H and the surface electric potential ψ_0:

$$\sigma_H = \frac{2\varepsilon k_B T_{abs} \kappa}{e} \sinh\left(\frac{e\psi_0}{2k_B T_{abs}} \right) \tag{2.16}$$

where ε represents the dielectric constant of the solution, κ is the inverse of Debye length, representing the thickness of the diffuse double layer, e stands for the elementary charge (1.602×10^{-19} C), and sinh(x) is the hyperbolic sine function, given by $(e^x - e^{-x})/2$.

The high concentration of Na^+ ions in solution at elevated alkalinity levels (e.g., pH > 9.6 for Case 1; Stage 2 in Figure 2.15) leads to a reduction in both the surface electric potential ψ_0 and the surface charge σ_H, as described by Eq. (2.16). This indicates a decrease in the repulsive forces of the electrical double layer between the matrix and SPs (Bai et al., 2025b). As a result, the attachment coefficient rises (Figure 2.14c), leading to a subsequent decrease in the recovery rate (Figure 2.14d). High alkalinity can lead to the formation of large aggregates, which may plug the pore throats due to particle mobilization, further hindering the transport of SPs, particularly at lower seepage velocity (e.g., $v = 0.066$ cm/s; Figure 2.14d). In contrast, at high seepage velocity (e.g., $v = 0.199$ cm/s), the recovery rate increases.

CHAPTER 3

Attachment–Detachment Model with Adsorption Hysteresis

3.1 MAIN CONTENT OVERVIEW

This chapter proposes a nonlinear attachment–detachment model with hysteresis, which employs an attachment-function along with scanning detachment isotherms to simulate the attachment/detachment of suspended substances, such as insoluble suspended particles (SPs) as well as soluble heavy metal ions (HMs). This model, referred to as the Bai model (Bai et al., 2021b), incorporates seepage velocity as a factor influencing attachment-history and temporal variations in SPs (or HMs) concentration, thereby offering improved theoretical solutions for suspended substance transport. This chapter examines the attachment mechanism of ionic, nanoscale, and microscale substances—spanning in size from ions to larger insoluble particles—onto porous media such as quartz sand. Temperature effects are incorporated into the analysis. Experimental tests under varying injection concentrations and seepage velocities are conducted to validate the model's reliability.

On the other hand, this chapter investigates the competitive adsorption of multiple HMs onto red mud particles (RMs), with particular emphasis on the influence of temperature. Building upon this, a further examination is conducted into the detachment characteristics behaviors of HMs previously adsorbed on RMs. These findings establish a foundational basic for developing purification technologies targeting HMs pollutants in aqueous systems or shallow groundwater.

3.2 A NEW NONLINEAR ATTACHMENT–DETACHMENT MODEL

3.2.1 Static Attachment–Detachment Model

Regarding the transient equilibrium attachment of SPs (or HMs, etc.), the concentration of attached particles on the porous matrix often exhibits a nonlinear relationship with the concentration of SPs in solution—a relationship that is clearly influenced by the

DOI: 10.1201/9781003783749-3

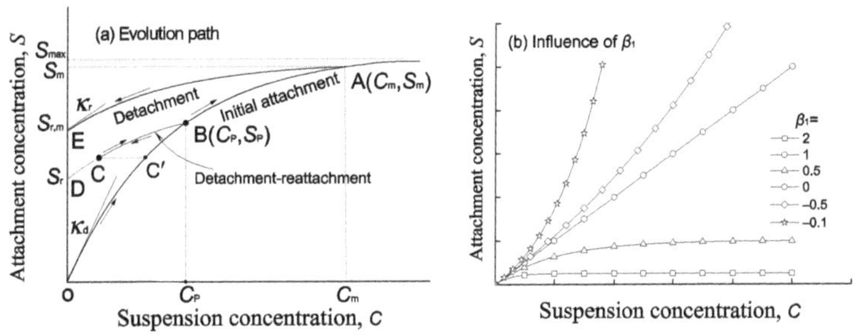

FIGURE 3.1 Model of attachment–detachment. (a) Evolution path. (b) Influence of β_1.

attachment-history of the porous media (Bai et al., 2019a, 2021a). For instance, a clean porous medium free of SPs attachment (point O in Figure 3.1a, $S = 0$). Here, the increasing rate at which the attached concentration S diminishes as the solution concentration C increases. Correspondingly, the attachment path progresses from point O toward point A, passing through point B, until reaching the maximum attainable concentration, S_{max} (saturation concentration). If the size of SPs falls below a certain threshold, significant blocking within the porous media does not occur; instead, the attachment quantity of the matrix declines with increasing C. Ultimately, the attached concentration approaches the theoretical matrix S_{max} when C reaches C_{m} (point A in Figure 3.1a). Hence, as $C \to \infty$, $S \to S_{\text{max}}$, which is referred to as the "initial attachment process" ($O \to B \to A$ in Figure 3.1a).

Deposited SPs on porous media can be remobilized when the suspension concentration, C, falls below C_{m} (Figure 3.1a). This remobilization is termed the detachment or release process ($A \to E$). As a result, the attached concentration decreases to the residual level, $S_{\text{r,m}}$, when C approaches zero. If the suspension concentration rises from zero (point O) to a specific value C_{P} (point B) and subsequently declines, the detachment follows a path from B to D via point C ($B \to C \to D$). This trajectory is described as a scanning detachment isotherm. The value C_{P} at point B is designated as the "previous concentration." These scanning curves occupy the region between the initial attachment curve (C_{P}) and the zero suspension concentration ($C = 0$). Should the suspension concentration decrease to an intermediate point (e.g., point C) and then increase again, attachment resumes along the path $C \to B$. This is called the "reattachment process." The shape of the reattachment path is assumed to be identical to that of the prior detachment segment $B \to C$. This behavior depends on the SPs concentration history during attachment or detachment and is characterized through static attachment tests. Nonetheless, if C continues to increase beyond C_{P}, the attachment process resumes along the original path from B to A. Similar dynamic patterns were reported by Van Genuchten and Wierenga (1977) for the transport of the herbicide 2,4,5-T (2,4,5-trichlorophenoxyacetic acid) through Glendale soil under static conditions.

A nonlinear static model for the attachment and detachment, which accounts for the attachment-history of the porous media, can be formulated. Under conditions where $\partial C / \partial t > 0$ and $C > C_{\text{P}}$ ($O \to B \to A$) in Figure 3.1a, the relationship between attached concentration S and the suspension concentration C can be expressed as follows (i.e., attachment-function):

$$S = S_{\max} \cdot (1 - e^{-\beta_1 \cdot \frac{C}{C_L}}) \tag{3.1}$$

where C_L is the characteristic concentration (whose dimension is the same as that of C), which is assumed to be $C_L = 1$, and β_1 is the dimensionless attenuation factor of the initial attachment process.

From Eq. (3.1),

$$\frac{\partial S}{\partial C} = \kappa_d \cdot e^{-\beta_1 \cdot \frac{C}{C_L}} \tag{3.2}$$

where $\kappa_d = \beta_1 \cdot S_{\max} / C_L$, κ_d is the transient equilibrium constant during the initial attachment process $[M^{-1}L^3]$.

When $C = 0$, $\partial S / \partial C = \kappa_d$; when $C \to \infty$, $\partial S / \partial C = 0$. When $\partial C / \partial t < 0$ or when $\partial C / \partial t > 0$ and $C < C_P$, the detachment and reattachment process ($B \to C \to D$ in Figure 3.1a) is described as follows:

$$\frac{\partial S}{\partial C} = \kappa_r \cdot e^{-\beta_2 \cdot \frac{C}{C_P}} \tag{3.3}$$

where κ_r is the transient equilibrium constant during the detachment process $[M^{-1}L^3]$ and β_2 is the dimensionless attenuation factor of the detachment process (i.e., scanning detachment isotherms).

The parameter C_P serves to distinguish between individual scanning curves. Larger values of β_1 and β_2 correspond to a more rapid approach to equilibrium, thereby indicating a lower attachment-capacity of the porous medium. In the specific case where $\beta_1 = \beta_2 = 0$, Eqs. (3.2) and (3.3) simplify to represent a linear attachment process. Furthermore, when $\beta_1 = \beta_2 = 0$ and condition $\kappa_d = \kappa_r$ holds, the system can be accurately described by the Freundlich linear attachment model.

From Eq. (3.2), the maximum attachment concentration can be written as $S_{\max} = \kappa_d C_L / \beta_1$. From Eq. (3.1), when $C = C_P$, the previous attachment concentration is

$$S_P = \frac{\kappa_d C_L}{\beta_1} \cdot (1 - e^{-\beta_1 \cdot \frac{C_P}{C_L}}) \tag{3.4}$$

In the detachment–reattachment process ($B \to C \to D$ in Figure 3.1a), $S = S_P$ when $C = C_P$. Based on this condition, Eq. (3.3) can be integrated to yield the relationship between the attachment concentration S and suspension concentration C during this process as follows:

$$S = \frac{\kappa_d C_L}{\beta_1} \cdot (1 - e^{-\beta_1 \cdot \frac{C_P}{C_L}}) - \frac{\kappa_r C_P}{\beta_2} \cdot (e^{-\beta_2 \cdot \frac{C}{C_P}} - e^{-\beta_2}) \tag{3.5}$$

For different previous concentrations of C_P, different scanning curves are obtained. Therefore, the residual concentration ($C = 0$, Figure 3.1) can be obtained:

$$S_r = \frac{\kappa_d C_L}{\beta_1} \cdot (1 - e^{-\beta_1 \cdot \frac{C_P}{C_L}}) - \frac{\kappa_r C_P}{\beta_2} \cdot (1 - e^{-\beta_2}) \tag{3.6}$$

When $C_P \to 0$, one has $S_r = 0$. When $C_P = C_m$, the residual concentration can be obtained:

$$S_{r,m} = \frac{\kappa_d C_L}{\beta_1} \cdot (1 - e^{-\beta_1 \cdot \frac{C_m}{C_L}}) - \frac{\kappa_r C_m}{\beta_2} \cdot (1 - e^{-\beta_2}) \tag{3.7}$$

where $S_{r,m}$ is the maximum residual concentration of SPs on the porous medium matrix, which is lower than the potential attachment concentration S_m (Figure 3.1a).

Different C_P values correspond to distinct yet similar detachment processes (i.e., scanning curves), which share the same detachment coefficient κ_r (i.e., release coefficient). The shape of the initial attachment process, however, can vary for different transported substances. From Figure 3.1b, a classical linear attachment model describes the attachment curve when $\beta_1 = 0$. When $\beta_1 < 0$, the curve exhibits an upward concavity, with the attachment rate increasing in proportion to the absolute value of β_1. This implies that the attachment rate accelerates as the solution concentration rises, a behavior often associated with noticeable straining effects. Such straining, typically observed for substances with larger particle sizes, can lead to a significant reduction in porosity within the porous medium. Conversely, when $\beta_1 > 0$, the attachment curve gradually flattens as concentration increases, theoretically approaching a maximum attachment quantity (S_{max} in Figure 3.1a). This limiting capacity is governed by the balance of forces acting on suspended substances in flowing water and their attachment torque.

3.2.2 Nonequilibrium Process during the Transport

The attachment of suspended substances is fundamentally a nonequilibrium hydrodynamic process. Effects such as size exclusion or preferential flow can further accentuate the nonequilibrium characteristics influencing the transport and attachment of suspended substances. If we denote the concentration of suspended substances in solution at time t as C and let the attachment concentration on the matrix under nonequilibrium conditions be σ. Then, the attachment rate (or the detachment rate) can be formulated as follows:

$$\frac{\partial \sigma}{\partial t} = \lambda \cdot (S - \sigma) \tag{3.8}$$

where λ is the reaction rate constant [T^{-1}].

The parameter λ quantifies the mass transfer between flowing and deposited SPs under hydraulic conditions. This mass exchange inherently exhibits hysteretic behavior in the attachment of solid particles to the porous medium matrix, a behavior that can be derived from the physicochemical properties of both the SPs (or solute) and the porous medium itself. The equilibrium attachment concentration S during the initial attachment phase (e.g., point C' in Figure 3.1) is obtained from Eq. (3.1), while the corresponding value during the equilibrium detachment–reattachment process (e.g., point C in Figure 3.1) is calculated using Eq. (3.5). Notably, the attachment rate at point C in the detachment–reattachment process equals the rate at the same attachment concentration (point C') in the initial attachment curve. When the concentration of SPs in the flowing water decreases ($B \rightarrow C \rightarrow D$), λ assumes a negative value, which physically corresponds to the detachment or release process.

From Eq. (3.8), it follows that $\partial\sigma / \partial t = f(S,\sigma)$, showing that the attachment rate of SPs depends not only on the solution concentration C, but also on the attached concentration σ on the matrix, the potential attachment-capacity S_m, and even the previous concentration C_p. When $\sigma = S$, $\partial\sigma / \partial t = 0$, meaning the attachment rate reaches zero. For both the initial attachment and the detachment–reattachment processes, the attachment rate (i.e., $\partial\sigma/\partial t$) attains its maximum value when $\sigma = 0$.

In the initial attachment stage, where time $t = 0$ and $\sigma = 0$, the evolving attachment concentration of SPs throughout the transport process can be derived from Eq. (3.8) as follows:

$$\sigma = S(1 - e^{-\lambda t}) \tag{3.9}$$

In the detachment–reattachment process, the initial condition at $t = 0$, $\sigma = S_r$. Accordingly, the attachment concentration of SPs during this process can be determined by Eq. (3.8):

$$\sigma = S(1 - e^{-\lambda t}) + S_r e^{-\lambda t} \tag{3.10}$$

The proposed attachment–detachment model comprises three unknown variables (C, S, and σ), which are solvable using Eq. (3.2) or Eq. (3.3) together with Eq. (3.8). This formulation ensures the completeness of the equation system.

3.3 EXPERIMENTS AND VERIFICATION

3.3.1 Static Attachment Test

An experimental apparatus was designed to determine the physical parameters relevant to the static attachment and detachment processes of SPs (Figure 3.2). The cylindrical test chamber, constructed from an organic material, has a height of $h = 20$ cm and a diameter of $\phi = 8$ cm. A 2-mm diameter water outlet is located at the bottom of the chamber. Positioned 5 cm above this outlet is a 0.5 mm stainless steel mesh screen. The specimen height within the chamber was set to 15 mm, representing a unit-scale experimental configuration.

The SPs solution of specified concentration (e.g., C_P in Figure 3.1a) was injected into the cylinder via a specially designed dispenser. During the injection, the base of this disperser was maintained 10 cm above the top surface of the specimen until the specimen was fully submerged in the suspension. Subsequently, the bottom outlet was opened to allow a

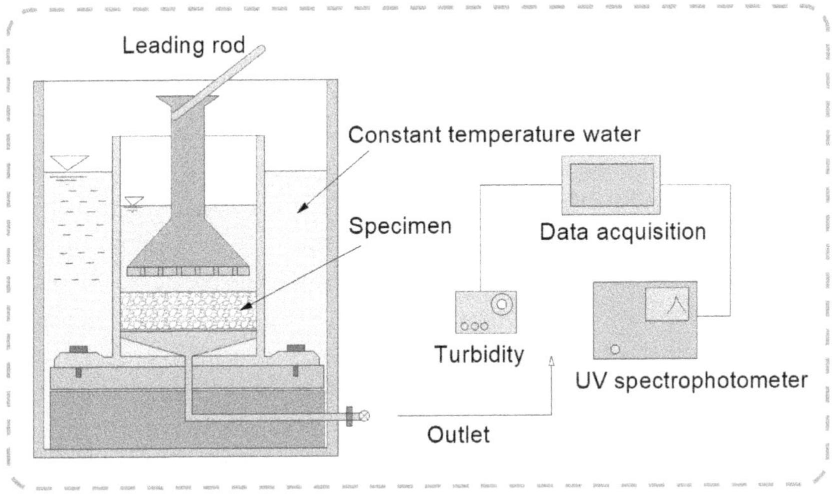

FIGURE 3.2 Test apparatus.

controlled, slow discharge of the suspension at a controlled seepage velocity of $v < 0.01$ cm/s, facilitating the static attachment of SPs within the porous medium. The installation screen served to retain the porous media while permitting the effluent to exit. Suspension concentration was monitored using a precalibrated correlation between turbidity and concentration. From this measurement, the corresponding attachment concentration within the porous media (e.g., S_p in Figure 3.1a) was calculated. Once a stable attachment was achieved, the suspension concentration was systematically lowered to induce a controlled detachment process (e.g., $B \rightarrow C \rightarrow D$ in Figure 3.1a). Similarly, other detachment scenarios (e.g., $A \rightarrow E$ in Figure 3.1a) could be investigated using an analogous procedure.

The used porous media was quartz sand, characterized by a particle size distribution of 26% within the 1–2 mm range and 74% within the 2–3.5 mm range. Its physical parameters include a uniformity coefficient of $C_u = 1.5$, a coefficient of curvature of $C_c = 1$, and a density of $\rho_s = 2.65\,g/cm^3$. The SPs consisted of spherical SiO_2 particles manufactured from natural silica sand, with a density of $\rho_s = 2.26\,g/cm^3$. Prior to testing, all materials were purified by washing in an acidic solution followed by rinsing with high-purity deionized water, and were subsequently dried at 105 °C for 24 h. The specimen was prepared with a porosity of $n = 0.42$. Static attachment tests were conducted using SPs with median particle sizes of $D_{50} = 2\,\mu m$, 10 μm, and 50 μm. All experiments were carried out at a controlled room temperature (22–24 °C).

3.3.2 Column Experiment

Column experiments were performed to discuss the transport behavior of the SPs, as outlined in Chapter 2. In these tests, the specimen column had a height of h = 30 cm and a diameter of ϕ = 8 cm, yielding an aspect ratio (h/ϕ) of 3.75. The column was packed in layers to a porosity of n = 0.42, consistent with the static attachment tests described in Section 3.3.1. Throughout the experiments, the water level was maintained at 1–2 cm above the soil surface to ensure full saturation. Flow was regulated using a peristaltic pump, which supplied high-purity deionized water from the feed tank. A 0.5-mm mesh screen was installed at both the inlet and outlet to promote uniform flow distribution into and out of the soil column.

The calibration of transport parameters is demonstrated here using SPs with a median particle size of D_{50}=10 μm. After the seepage velocity stabilized, a pulse-injection of SPs solution was introduced into the soil column. This involved instantaneously injecting a volume of V_{inj}= 30 ml over a very short duration within $t_{\text{inj}} = 2$ s (i.e., pulse-injection pattern). The tests were conducted under three different seepage velocities (v = 0.087, 0.173, and 0.260 cm/s) and four injection concentrations ($C_{\text{inj}} = 0.5, 1, 3, \text{and } 4 \text{ mg/ml}$).

To further validate the reliability of the proposed attachment–detachment model, additional transport experiments were performed using a continuous injection scheme with variable injection concentration and seepage velocity. For particle sizes (i.e., $D_{50} = 2 \text{ or } 10\,\mu\text{m}$), two variable-concentration patterns were tested: one where the injection concentration was sequentially increased from $C_{\text{inj}} = 0.1$ mg/ml to 0.5 mg/ml and then to 2 mg/ml at set time intervals under three seepage velocities (v = 0.087, 0.173, and 0.260 cm/s); and another where $C_{\text{inj}} = 2$ mg/ml is decreased to 0.5 mg/ml and then to 0.1 mg/ml at a fixed velocity of v = 0.260 cm/s. In addition, four variable-velocity patterns were examined for three particle sizes ($D_{50} = 2, 10, \text{and } 50\,\mu\text{m}$) at a constant $C_{\text{inj}} = 2$ mg/ml. These included three scenarios of increasing seepage velocity (v = 0.087→0.173 cm/s, v = 0.087→0.260 cm/s, and v = 0.173→0.260 cm/s) and one scenario of decreasing velocity (v = 0.260→0.173 cm/s).

3.3.3 Parameter Sensitivity Analysis

3.3.3.1 Coordinate Transformation

A one-dimensional scenario is examined, with the initial concentration of SPs assumed to be zero. The penetration rate M_{b} (%) is expressed as the ratio of the mass of SPs that have passed through a given cross section x by time t to the total mass of SPs introduced into the porous medium:

$$M_{\text{b}} = \frac{\int_0^t C(x,t)\cdot u\cdot(nA)\cdot \mathrm{d}t}{m} \tag{3.11}$$

The M_{b} is also called recovery rate in Chapter 2. The definition $M_{\text{a}} = 1 - M_{\text{b}}(\%)$ is the amount of SPs deposited in the range of (0, x). The dimensionless time is defined as $t_{\text{p}} = (unAt)/(nAL) = ut/L$. Thus, Eq. (3.11) becomes

$$M_{\text{b}} = \frac{nAL\cdot\int_0^{t_{\text{p}}} \tilde{C}\left(x,t_{\text{p}}\right)\cdot \mathrm{d}t_{\text{p}}}{m} \tag{3.12}$$

where $\tilde{C}(x,t_{\text{p}})$ represents the dimensionless concentration, L is the column length, t_{p} represents the ratio of the water volume $Q_{\text{t}}(Q_{\text{t}} = unAt)$ passing through the cross section x over time t to the total void volume of the column ($V_0 = nAL$). This ratio is equivalently expressed as the number of pore volume $V_P(V_P = t_p)$ (Bai et al., 2019a). Differentiating Eq. (3.12) with respect to t_p given the relative concentration C_R:

$$C_R = \frac{V_0}{m} \cdot \tilde{C}(x, t_p) \tag{3.13}$$

3.3.3.2 Parameter Sensitivity

When the curvature of the initial attachment path ($O \to B \to A$ in Figure 3.1a) closely resembles that of the detachment–reattachment process ($B \to C \to D$ in Figure 3.1a), it can be approximated that $\beta = \beta_1 = \beta_2$. Subsequently, the ratio $\eta = \kappa_r / \kappa_d$ is defined to quantify the influence of the detachment effect on attachment. For harmonic variations in SPs concentration at the column surface, the relative concentration C_R in the effluent is influenced by three parameters within the attachment–detachment model: β, η, and λ. The temporal variation of the injected SPs concentration is assumed to follow the form shown in Figure 3.3.

$$C = C_0 \cdot \sin\left(\frac{2\pi}{T} t\right) \tag{3.14}$$

where C_0 is the concentration amplitude and T is the period.

Here, we set $C_0 = 2$ mg/ml, $T = 120$ s (i.e., the injection time is 60 s), and an injection velocity of $v = 0.087$ cm/s. Figure 3.3 illustrates the sensitivity of the parameters β, η, and λ on the recovery rate M_b of SPs in the effluent, where C_{out} denotes the outflow concentration. The calculations indicate an M_b value of 82.6% for $\beta = 0.01$, $\eta = 0.5$ $(\kappa_d = 0.01\,\text{ml/mg})$ and $\lambda = 0.02 \times 10^{-3}$. As the attenuation factor β rises from 0 to 1, the penetration rate M_b initially increases and then stabilizes. This demonstrates that a nonlinear attachment process ($\beta > 0$) can reduce the overall amount of SPs attachment. At $\beta = 1$, the penetration rate reaches 98.7%, reflecting a pronounced decline in deposited mass.

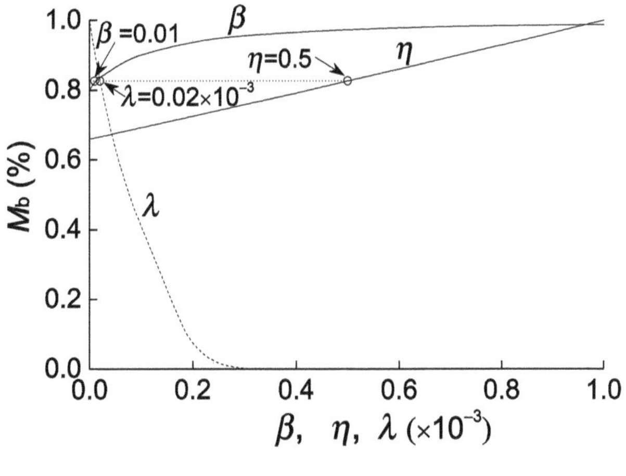

FIGURE 3.3 Parameter sensitivity for the attachment–detachment process.

As the value of η ($\eta = \kappa_r / \kappa_d$, $\eta =$ 0-1) increases, the M_b rises in an approximately linear manner, reaching $M_b =$ 100% when $\eta = 1$. Figure 3.3 indicates that as the injection concentration of SPs declines (i.e., after $C_{out} / C_0 = 1$), the detachment effect becomes more pronounced with higher η values (e.g., η = 0.2→0.5→0.9), resulting in a more distinct tailing phenomenon in the BTCs. The reaction rate constant λ reflects the intensity of interaction between SPs and the porous matrix. As λ increases, the penetration rate M_b drops sharply from $M_b =$ 100% in the absence of attachment (λ = 0) to $M_b = 0$ when SPs and the matrix are fully coupled ($\lambda = 0.3\times10^{-3}$). Therefore, λ significantly influences the modeled outcomes, with its typical range (0–0.3×10^{-3}) being closely tied to the physicochemical properties of both the SPs and the porous medium surface.

3.3.3.3 Gaussian Pulse-Injection

To assess the applicability of the attachment–detachment model, the seepage velocity is described as decreasing exponentially over time:

$$u(t) = u_i \cdot e^{-\alpha t} \tag{3.15}$$

where u_i is the initial seepage velocity and α is the attenuation parameter [T^{-1}].

A constant seepage velocity corresponds to the case where α = 0. The injection concentration of SPs at the column surface, $C_0(0, t) = f(t)$, is assumed to follow a pulse periodic pattern (Figure 3.4). This pattern is represented by the Gaussian function:

$$f(t) = \begin{cases} \dfrac{W}{\sqrt{2\pi}} \cdot e^{-\frac{[(t-t_{i,N})-b]^2}{2}} & , t_{i,N} \le t \le t_{i,N} + 2b \\ 0 & , \text{otherwise} \end{cases} \tag{3.16}$$

where $t_{i,N}$ denotes the starting time of pulse N, N represents the cycle number, b is the peak characteristic parameter of the Gaussian function, and W indicates the injection mass per pulse. Consequently, $W = \int_{t_{i,N}}^{t_{i,N}+2b} f(t)\,dt$.

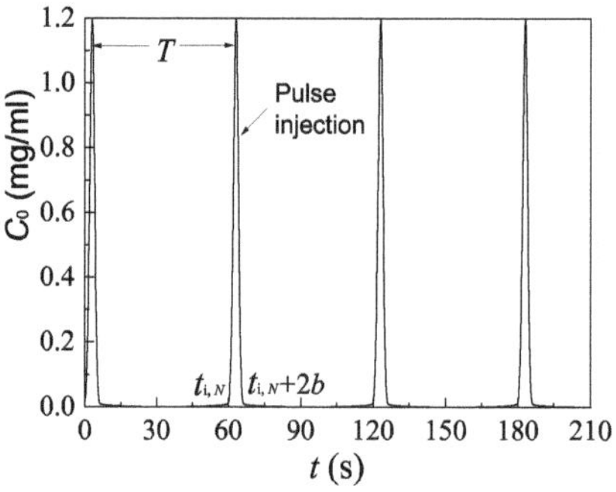

FIGURE 3.4 Pulse-injection pattern.

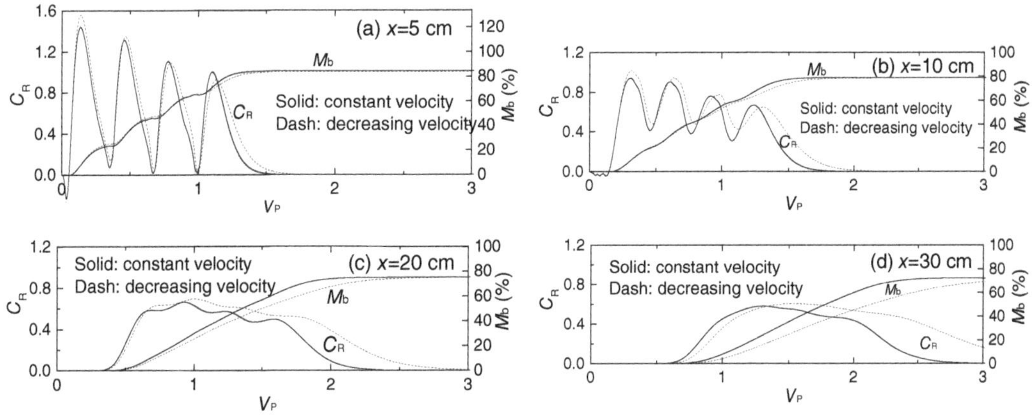

FIGURE 3.5 Penetration process at (a) x = 5 cm, (b) x = 10 cm, (c) x = 20 cm, and (d) x = 30 cm.

Assuming each pulse-injection introduces $V_{\mathrm{inj}} = 30\,\mathrm{ml}$ of SPs over a duration of $t_{inj} = 2b = 2\ s$. The initial seepage velocity in Eq. (3.15) is set as u_{i}=0.15 cm/s. Other defined parameters include $\alpha_{\mathrm{d}} = 0.4\,\mathrm{cm}$, $\beta = \beta_1 = \beta_2$, and a velocity attenuation coefficient of $\alpha =\ 0.001\ \mathrm{s}^{-1}$. Figure 3.5 displays the temporal variation in the relative concentration of SPs in the fluid at different distances from the injection surface (x = 5, 10, 20, and 30 cm). For this simulation, the injection period is T = 60 s, the attenuation factor β = 0.01, the reaction rate constant $\lambda =\ 2\times10^{-5}\,\mathrm{s}^{-1}$, the transient equilibrium constant κ_{d}= 0.01 ml/mg and κ_{r}= 0.005 ml/mg. At $x = 5$ cm, the relative concentration of SPs shows periodic variation, with peak amplitudes diminishing as pore volume (V_{P}) increases. This indicates that interactions between SPs in the flowing water and the porous matrix intensify over time, leading to greater attachment. However, with increasing distance from the injection point, the periodic signal becomes less distinct (e.g., x = 20 cm; Figure 3.5c) and eventually vanishes (e.g., x = 30 cm; Figure 3.5d). This attenuation is attributed to the combined effect of convection, dispersion, and attachment during transport.

Under constant velocity conditions (α = 0, see the solid curves in Figure 3.5), the attachment mass M_{a} reaches 16.6% (corresponding to a penetration rate $M_{\mathrm{b}} = 84.4\%$) within the interval x = 0–5 cm and increases to $M_{\mathrm{a}} = 27.5\%$ ($M_{\mathrm{b}} = 72.5\%$) over the full column length (x = 0–30 cm). Consequently, the attachment mass within the x = 5–30 cm interval represents merely 10.9%, demonstrating a rapid decline in attachment with increasing distance from the injection surface. Figure 3.5 reveals that at x = 5 cm, the SPs concentration becomes very low once the V_{P} exceeds 1.5. When the distance extends to x = 20 cm, the relative concentration drops to approximately zero at around $V_{\mathrm{P}} = 2.5$, indicating a pronounced retardation effect. Moreover, the peak relative concentration diminishes sharply with transport distance; for the first peak, $C_{\mathrm{R}} = 1.45$ at x = 5 cm, but declines to $C_{\mathrm{R}} = 0.58$ at x = 30 cm. In scenarios where the seepage velocity decays (α = 0.001 s^{-1}), the peak concentration C_R initially exceeds that of the nonattenuated case (α = 0) before gradually diminishing with successive cycles. Furthermore, as the distance x increases, the occurrence of the concentration peak is delayed and the overall penetration rate M_b decreases, a consequence of the reduced seepage velocity in the later stages of the experiment.

The simulation results demonstrate that for the nonlinear model described by Eqs. (3.2) and (3.3), where $\kappa_d = 0.08$ ml/mg, $\kappa_r = 0.04$ ml/mg, $\beta = 0.01$ and its simplified linear form ($\beta = 0$), the penetration rate remains below 100% following cyclic SPs attachment. Variations in the pulse-injection period led to marked differences in the temporal concentration profile, peak effluent concentration, and the total mass of SPs retained during transport. For example, at x = 20 cm and with β = 0.01, the penetration rates are $M_b = 82.2\%$, 72.1%, and 46.9% for periods of T = 6, 60, and 120 s, respectively. Furthermore, a shorter injection period (e.g., T = 6 s) corresponds to a more rapid rise in penetration rate and a higher peak relative concentration C_R. This behavior is linked to the reaction rate constant λ between the SPs and the porous matrix, reflecting the nonequilibrium nature of the hydrodynamic process. The analysis also indicates that increasing the nonlinear parameter β (e.g., β = 0→0.01→0.1 at T = 60 s) raises the penetration rate ($M_b = 63.7\% \to$ 72.1%→74.4%). This suggests that employing a linear attachment model may overestimate the actual attachment-capacity of the porous medium.

3.3.4 Experimental Calibration of the Transport Parameters

Figure 3.6 illustrates the variation in the static attachment concentration S with increasing or decreasing suspension concentration (C) of SPs. The rate of increase tends to diminish as C rises. For small particle sizes of D_{50} = 2 μm, observable attachment dominates, whereas for larger sizes like D_{50} = 50 μm, a straining effect becomes prominent. Consequently, the medium-sized particles exhibit relatively lower attachment amounts. Notably, the detachment process of SPs initially progresses gently before becoming steeper, with the final

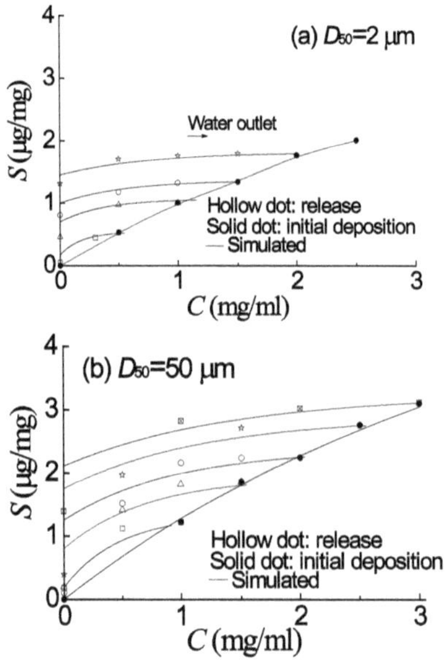

FIGURE 3.6 Static attachment tests. (a) D_{50} = 2μm. (b) D_{50} = 50μm.

deposited mass varying significantly across different particle sizes. The static attachment curve was fitted using Eq. (3.2) to derive parameter β_1. The detachment process was fitted using Eq. (3.3), under the assumption that parameter β_2 remains constant across all detachment paths (i.e., scanning detachment isotherms), thereby establishing the relationship between S and C/C_P for every detachment process.

Figure 3.7 presents the experimental BTCs for SPs (D_{50} = 10 μm; $C_{inj} = 0.5$ mg/ml) under pulse-injection conditions, along with the corresponding simulated BTCs fitted using COMSOL Multiphysics (COMSOL Co., Ltd.). The computational settings were defined as follows: a temporal discretization step of 3 s, a spatial discretization step of 0.02 m, a damping factor of 0.9, 4 iterations, and a relative tolerance of 0.0001. The model involves six key parameters (β_1, β_2, κ_d, κ_r, α_d, and λ). Among these, β_1 and β_2 are derived from static attachment tests. In general, dispersivity α_d tends to increase with both particle size and seepage velocity, but shows little dependence on injection concentration. Based on the experimental results, α_d is taken as 0.4 cm. The remaining three parameters κ_d, κ_r, and λ are uniquely calibrated by fitting the pulse-injection experimental data shown in Figure 3.7. The iterative fitting procedure continues until the specified accuracy is achieved.

For the case where D_{50} = 10 μm (Figure 3.7), the determination coefficient R^2 exceeds 0.99, indicating that the nonlinear attachment–detachment model exhibits strong agreement with the experimental data. The peak concentrations in the BTCs consistently appear after one pore volume (e.g., $V_P > 1$), a delay attributed to adsorption hysteresis (i.e., "adsorption lag" effect). Notably, the latter segments of the curves (e.g., $V_P > 2$) effectively illustrate the release of previously deposited SPs, as evidenced by effluent concentrations remaining above zero for an extended duration.

As the injection concentration rises, the penetration rate of SPs initially declines before subsequently increasing. For instance, at a seepage velocity of v = 0.260 cm/s, the penetration rate $M_b = 65.8\%$, 58.3%, 70.0%, and 76.6% for injection concentrations of $C_{inj} = 0.5$, 1, 3, and 4 mg/ml, respectively. Comparable trends are observed at seepage velocities of v = 0.087 and 0.173 cm/s. This behavior reflects a nonlinear relationship between the attached and suspension concentrations, confirming the presence of maximum attachment quantity

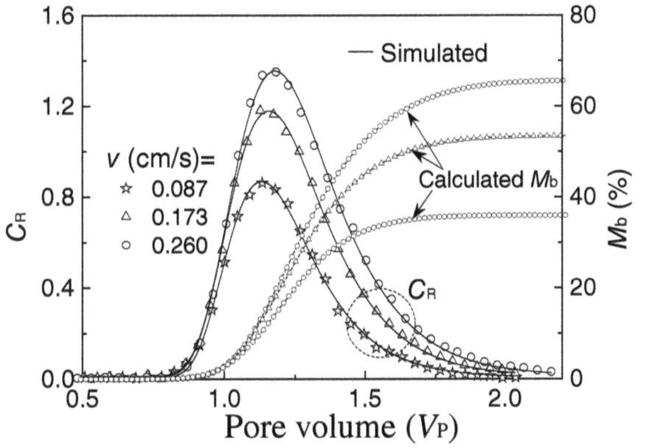

FIGURE 3.7 Pulse-injection and optimization-fitting results.

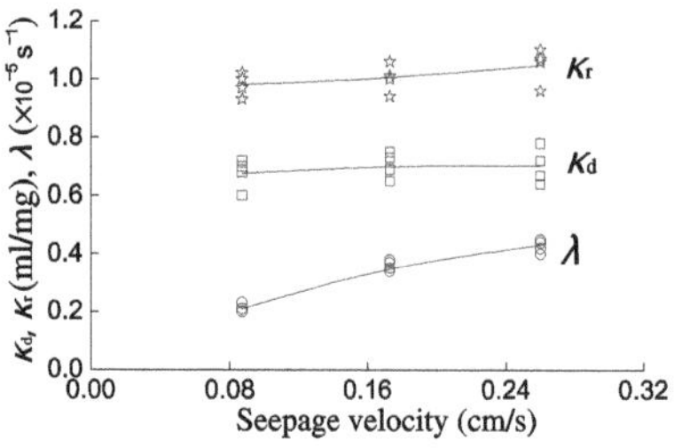

FIGURE 3.8 Transport parameters with seepage velocity.

σ_{max}. At lower suspension concentration, the dynamic attachment concentration σ on the matrix remains below saturation, allowing further attachment of SPs. Consequently, the penetration rate M_b decreases with increasing injection concentration. Once the suspension concentration exceeds a certain threshold, σ approaches saturation, and the increment in attachment concentration diminishes toward zero. Under these conditions, the porous matrix cannot retain additional SPs even at elevated injection concentrations, leading to an increase in the penetration rate.

The transport parameters (β_1, β_2, κ_d, κ_r, α_d and λ) governing SPs behavior in porous media are influenced by multiple factors, including seepage velocity, injection concentration, and particle size. The situation is further complicated by the interconnected nature of these factors and their continuous evolution over transport distance and time. As illustrated in Figure 3.8, the fitting values of κ_d, κ_r, and λ, derived from the data in Figure 3.7, vary as functions of seepage velocity under different injection concentrations ($C_{inj} = 0.5$, 1, 3, and 4 mg/ml). The results indicate that across the velocity range of v = 0.087–0.263 cm/s, κ_d and κ_r remain largely constant and can be treated as velocity independent. These parameters primarily characterize attachment under static conditions, as they relate more strongly to particle size and solution properties than to hydrodynamic velocity. Therefore, κ_d and κ_r can be taken as the average of their fitted values. For instance, for D_{50} = 10 μm, κ_d = 0.69 ml/mg and $\kappa_r = 1.01$ ml/mg. In contrast, Figure 3.8 shows that λ increases with seepage velocity, reflecting a stronger interaction between SPs and the porous matrix at higher flow. This demonstrates that seepage velocity plays a discernible role in modulating the transport dynamics of SPs.

3.3.5 Attachment–Detachment Mechanism of Ionic/Nanoscale/Microscale Substances

3.3.5.1 Materials

The quartz sand employed in the experiments consisted of 26% particles with diameters of 1–2 mm and 74% within the 2–3.5 mm range. This material exhibited a high degree of uniformity, characterized by a nonuniformity coefficient $C_u = 1.5$, a curvature coefficient $C_C = 1$, and a particle density of $\rho_s = 2.65$ g/cm^3. Given the relatively small size of SPs (<50 μm) and the properties of the quartz sand medium, significant straining effects could be effectively avoided during the attachment/detachment tests.

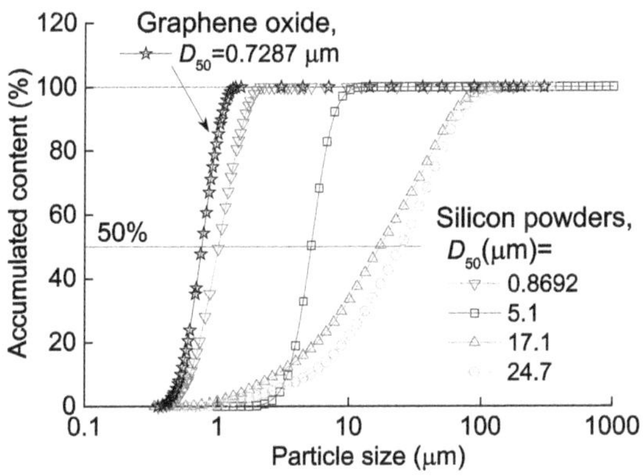

FIGURE 3.9 The PSDs of the selected SPs.

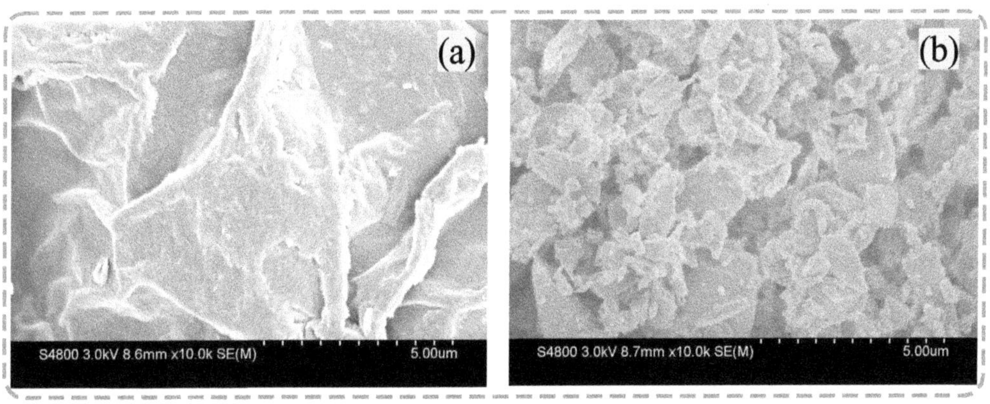

FIGURE 3.10 SEM images. (a) GO. (b) Silica powders.

Six different materials were selected for the static attachment–detachment experiments: copper ions (Cu^{2+}), GO, and four kinds of silica powders. Aqueous Cu^{2+} solutions were prepared using anhydrous copper nitrate ($Cu(NO_3)_2$) crystals, which appear as blue rhombic flakes, exhibit low toxicity, and possess high solubility in water. The GO sample had a particle size ranging from 0.3 μm to 2 μm. Its particle size distribution (PSD), as measured by laser diffraction, is shown in Figure 3.9. The median particle size D_{50} = 0.7287 μm, while dynamic light scattering yielded a median hydrodynamic diameter of 0.113 μm. Figure 3.10a displays a SEM image of the GO microstructure. A stable GO suspension with inert chemical properties was prepared, ensuring no reaction with the quartz sand in subsequent attachment–detachment tests.

Four types of irregular silica powders were derived from natural quartz sand, which is predominantly composed of SiO_2, with a particle density of $\rho_s = 2.26\,g/cm^3$. These materials were characterized by median particle sizes of D_{50} = 0.8692, 5.1, 17.1, and 24.7 μm. Further information regarding their PSD is illustrated in Figure 3.9. A representative SEM image corresponding to the sample with D_{50} = 0.8692 μm is also presented (Figure 3.10b).

The specific surface areas of the silica powders were determined using the nitrogen adsorption technique. Fourier transform infrared spectroscopy was employed to analyze the reaction spectra of GO, particularly focusing on its oxygen-containing functional groups. Additionally, the zeta potential distributions of both GO and the silica powders in aqueous suspension were assessed via nanoparticle size and zeta potential analysis.

For quantifying the concentration of the Cu^{2+} solution through ultraviolet spectrophotometry, a specific reagent was introduced to form a colored complex with Cu^{2+}. The absorbance of the solution was measured immediately after complex formation, and the corresponding Cu^{2+} concentration was derived using a pre-established calibration curve. Sodium diethyldithiocarbamate (DDTC-Na) was selected as the complexing agent. This white crystalline reagent produces a yellow-brown colloidal complex with Cu^{2+}, exhibiting a maximum absorption wavelength at 452 nm.

3.3.5.2 Copper Ions (Cu^{2+})

Figure 3.11 presents the experimental results of the static attachment/detachment process of Cu^{2+} on a SiO_2 porous medium. The optimal fitting curves in Figure 3.11a were obtained using the nonlinear model previously proposed by the author (i.e., Eqs. (3.1) and (3.5)). All fitting calculations were performed with Origin 2016 software (OriginLab Corporation, Northampton, USA). The fitted attachment equilibrium parameter κ_d was 0.086 ml/mg, with a very small attenuation factor (i.e., $\beta_1 = 0.0016$). Furthermore, the equilibrium detachment curves (i.e., scan curves) are nearly parallel, yielding an average κ_r value of 0.046 ml/mg and an attenuation factor $\beta_2 = 0.0022$. Given its positive charge, Cu^{2+} primarily adsorbs onto the negatively charged quartz sand surface via metal bonding. The electrostatic attachment effect is strong, while gravitational effects are negligible. From Figure 3.11b, the attachment-capacity of Cu^{2+} also increases with temperature. This is attributed to the enhanced irregular thermodynamic motion of Cu^{2+} at higher temperatures, which correspondingly increases the attachment/detachment equilibrium parameters.

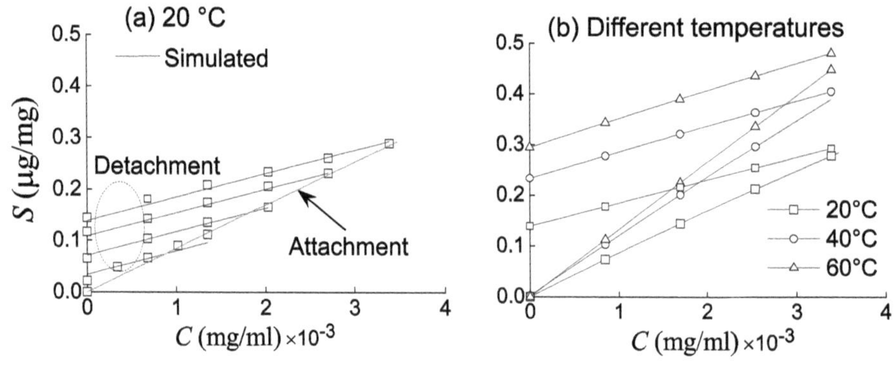

FIGURE 3.11 Attachment–detachment process for Cu^{2+}. (a) $T = 20$ °C. (b) Different temperatures.

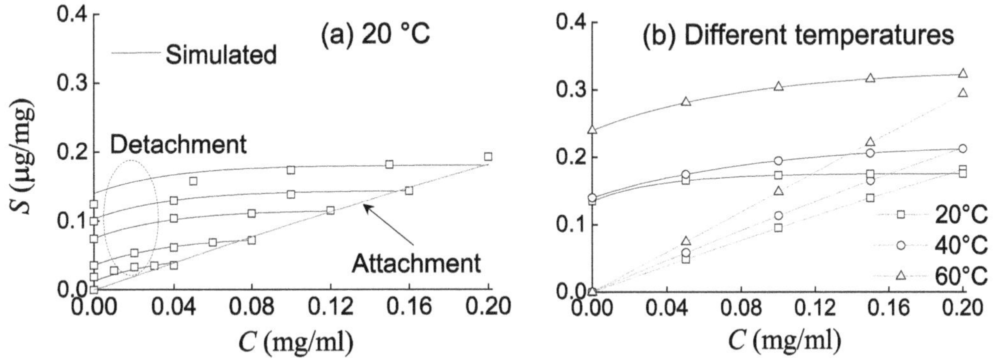

FIGURE 3.12 Attachment-detachment process for GO. (a) $T = 20$ °C. (b) Different temperatures.

3.3.5.3 *Graphene Oxide (GO)*

As illustrated in Figure 3.12, the particle size of GO (D_{50} = 0.7287 μm) is comparable to that of silica powders (D_{50} = 0.8692 μm, Figure 3.9). Their attachment/detachment behavior on the SiO_2 porous medium also demonstrated comparable trends (Figure 3.13). The attachment of GO displayed a strong, nearly linear profile, while during detachment the rate of change started slowly before gradually accelerating. Thus, the attachment/detachment mechanism of GO resembles that of fine silica powders and can be regarded as a predominantly physical attachment process on SiO_2. Fitting results produced a coefficient of determination $R^2 > 0.97$ for the attachment–detachment model, with only minor exceptions. The attenuation factor during the attachment was very small ($\beta_1 = 0.0010$, that is, almost no attenuation), whereas during the detachment it was $\beta_2 = 0.027$.

The experimental data indicate that the attachment-capacity of GO rises with increasing temperature (Figure 3.12), and the corresponding residual amount of retained GO also increases. This trend can be attributed to the enhanced thermodynamic energy of GO particles suspended in water at higher temperatures, which promotes more frequent collisions between GO and the porous matrix during the attachment phase, thereby increasing the likelihood of adsorption onto the surface. The attachment equilibrium parameter of GO showed a slight increase with temperature, while the detachment equilibrium parameter remained largely unaffected.

3.3.5.4 *Silica Powders*

Attachment–detachment experiments were conducted with four silica powders having median particle sizes (D_{50} = 0.8692, 5.1, 17.1, and 24.7 μm), tested at three temperatures (T = 20 °C, 40 °C, and 60 °C; Figures 3.13 and 3.14). Figure 3.13 indicates that during the attachment process, the amount retained on the porous matrix initially rose with increasing suspension concentration (C). For larger particles (e.g., D_{50} = 24.7 μm), the rate of increase slowed once the concentration exceeded a threshold, for example, $C > 1$ mg/ml (Figure 3.13). Essentially, as C increased, the attached amount gradually approached a maximum capacity, reflecting a weakening of further attachment and a more dominant role of bed filtration. In the detachment stage, the decrease in attached mass was initially

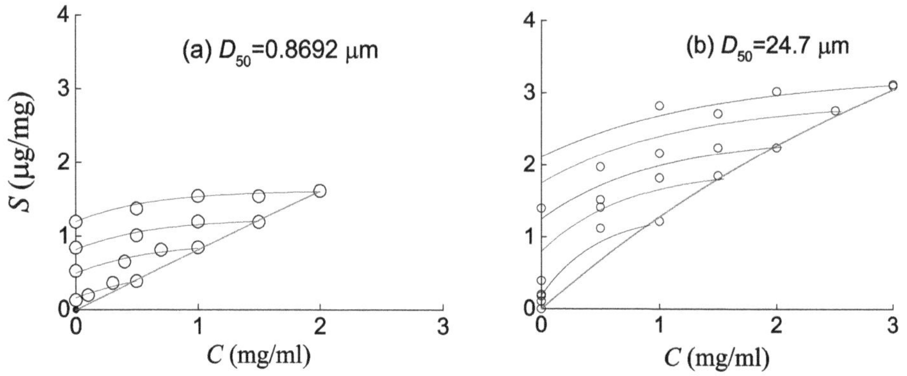

FIGURE 3.13 Attachment–detachment process for different silica powders. (a) D_{50} = 0.8692 μm. (b) D_{50} = 24.7 μm.

subtle as the suspension concentration was lowered. However, with further reduction in C, the equilibrium between particles in suspension and those on the solid matrix was progressively disturbed, leading to an accelerated detachment of particles from the surface. Naturally, the attachment behavior of the silica powders was also influenced by the hydrophobic characterization of silica powders and solid matrix, as well as interphase interactions at varying concentrations. Notably, the detachment curves for all four silica powders exhibited distinct nonlinear profiles.

Adjusting the temperature was found to directly influence the attachment and detachment behavior of silica powders (Figure 3.13; $T = 20^{\circ}\mathrm{C}$). As clearly illustrated in Figure 3.14, elevating the temperature enhanced the retention capacity of the porous media toward silica powders. Additionally, the peak attachment concentration shifted, and the attachment profiles exhibited a steeper trend, particularly for finer silica powders (e.g., D_{50} = 0.8692, 5.1 μm). This phenomenon can be interpreted as follows: a rise in temperature increased the internal energy of the silica powders, leading to more frequent collisions among silica powders themselves and with the porous matrix, thereby promoting their erratic motion. Naturally, for larger silica powders (e.g., D_{50} = 24.7 μm), this enhancing effect was considerably less pronounced. Conversely, the detachment rate demonstrated a positive correlation with increasing temperature (Figure 3.14), meaning the detachment curve grew steeper. In other words, during detachment, the equilibrium state of silica powders adhered to the matrix surface was more readily disrupted, resulting in a marked increase in the quantity of detached silica powders. Overall, the residual attachment amount (i.e., when the suspension concentration dropped to zero) rose with temperature for all particle sizes examined. Furthermore, as the temperature increased, the attachment/detachment equilibrium parameters corresponding to a specific particle size also increased, allowing a relationship between the equilibrium parameters and temperature to be established.

As illustrated in Figure 3.13, the attachment quantity (S) rose correspondingly with larger particle size within the finer size range (e.g., D_{50} < 5.1 μm). Nevertheless, at a particle size of D_{50} = 17.1 μm, the attachment-capacity initially declined before increasing again for a larger size (e.g., D_{50} = 24.7 μm). This pattern remained consistent across the three tested

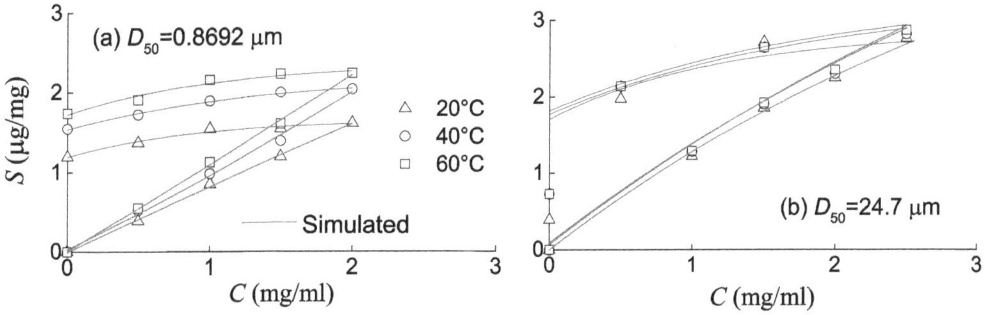

FIGURE 3.14 Temperature influence on silica powders. (a) D_{50} = 0.8692 μm. (b) D_{50} = 24.7 μm

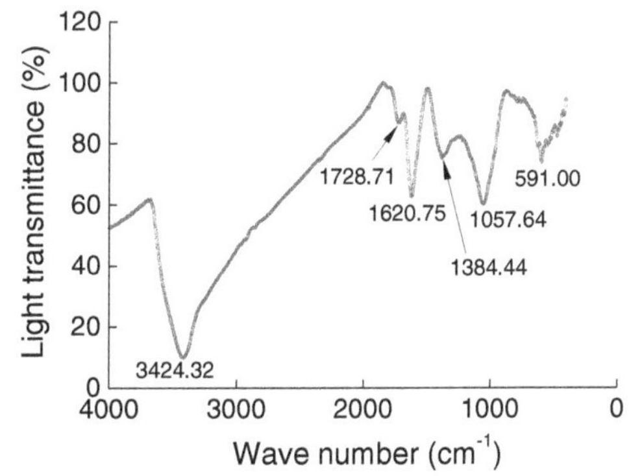

FIGURE 3.15 Functional group characteristics of GO.

temperatures (T = 20 °C, 40 °C, and 60°C; see Figure 3.14). In reality, the retention of silica powders within the soil column was governed by a complex interplay of adsorption and straining mechanisms. For smaller silica powders, adsorption played the primary role, while the porous medium demonstrated only a minor straining effect. With an increase in silica powder size, attachment onto the surfaces of the porous medium occurred more readily. However, upon reaching a sufficiently large size (e.g., D_{50} = 17.1 μm), the combined influence of downward gravity and size exclusion promoted transport along pore central pathways, thereby reducing surface attachment and the overall retained amount. Conversely, with a further increase in particle size (e.g., D_{50} = 24.7 μm) surpassing a certain pore dimension, the blocking effect became predominant. Consequently, the attachment equilibrium parameters for the silica powders were substantially higher than those for Cu^{2+}. It is important to note that the silica powders markedly decreased the porosity of the porous media, even causing the attachment curve to inflect upward (Figure 3.1b).

3.3.5.5 Microstructural Characterization

The GO is produced through the oxidation of graphite using strong acids and oxidizing agents. Its surface is abundant with oxygen-containing functional groups, which impart a strongly negative charge to the carbon layers. As depicted in Figure 3.15, GO exhibits distinctive peaks corresponding to high concentrations of hydroxyl (–OH; 3424.32 cm^{-1}, with low light transmittance), carbonyl (C=O, wave number 1728.71 cm^{-1}), and carboxylate (C—O—O; wave number 1057.64 cm^{-1}) groups. In addition, *a* > C=C < (alkene) stretching vibration, COO— asymmetric stretching vibration (wave number 1620.75 cm^{-1}), O—H in-plane deformation, and COO— symmetric stretching vibrations (wave number 1384.44 cm^{-1}), C—O—O stretching vibration, and O—H out-of-plane deformation vibration (wave number 591.00 cm^{-1}) were observed.

The specific surface area measurements for silica powders with particle sizes $D_{50} = 0.8692$, 5.1, and 24.7 μm were determined to be 6.571, 5.698, and 2.003 m^2/g, respectively. The specific surface area of GO reflects its dispersion state and lattice characteristics. Owing to its distinctive honeycomb structure (Figure 3.10a), GO possesses a comparatively large specific surface area (7.345 m^2/g) when contrasted with silica powders (Figure 3.10b). Nonetheless, the attachment interaction between GO and the negatively charged quartz sand matrix remained relatively weak. Images from the static attachment tests, obtained via metallographic microscopy at 5× magnification, depict SM adhered to quartz sand, as shown in Figure 3.16. For experimental clarity, the injected concentrations were set at $C = 0.5$ mg/ml for GO and $C = 5$ mg/ml for the silica powders in these tests.

As shown in Figure 3.16, GO primarily adhered to the matrix surface through an embedding mechanism (Figure 3.16a), displaying relatively weak adsorption. In contrast, silica powders demonstrated a progressively more pronounced embedding effect as particle

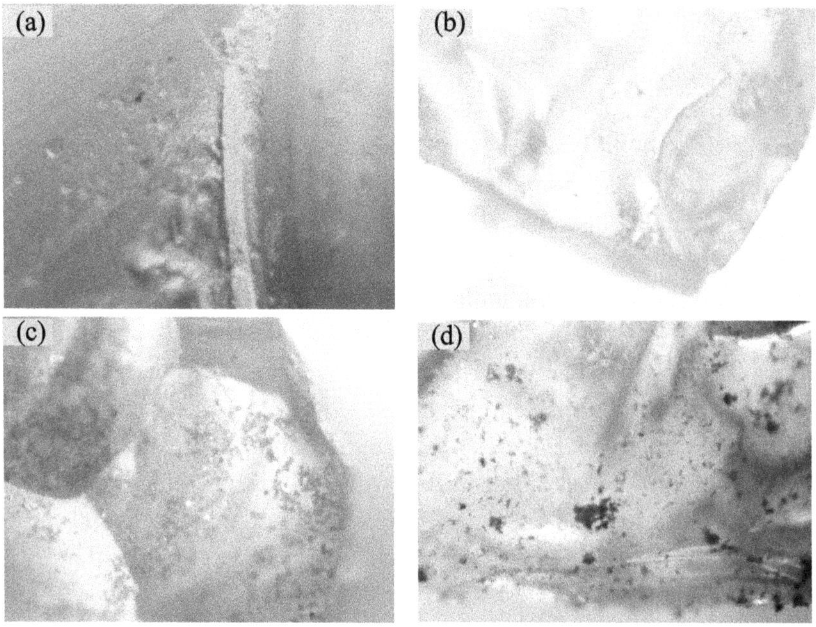

FIGURE 3.16 Metallographic microscopy images showing the attachment of GO and silica powders to quartz sand. (a) GO and silica powders with (b) $D_{50} = 0.8692$ μm, (c) $D_{50} = 5.1$ μm, and (d) $D_{50} = 24.7$ μm.

size increased (from Figure 3.16b to Figure 3.16d). This finding indicates that when GO serves as a purifying agent for negatively charged HMs, its migration distance—and thus its spatial influence range—exceeds that of commonly used positively charged SPs such as silica powders. Such extended mobility is often desirable in soil purification technologies. Furthermore, the honeycomb structure of GO is believed to exert a strong complexation effect on heavy metal pollutants, a subject reserved for future investigation.

The equilibrium attachment parameter κ_d of GO was marginally greater than that of silica powders of comparable particle size (i.e., D_{50} = 0.8692 μm). This suggests a somewhat stronger interfacial affinity of GO and the SiO_2 porous material, attributable to the functional groups present on GO (Figure 3.15). Furthermore, the detachment attenuation factor β_2 for GO (e.g., $\beta_2 = 0.027$ at $T = 20^{\circ}C$) was significantly lower than the corresponding values for silica powders (e.g., $\beta_2 = 1.61$ when $D_{50} = 0.8692$ μm and $T = 20^{\circ}C$). This outcome indicates that the detachment rate of GO was reduced relative to silica powders of the same particle size. In practice, the interaction force between GO and SiO_2 quartz sand is considerably stronger, as this interaction is largely governed by the oxygen-containing functional groups on their respective surfaces.

3.3.5.6 Zeta Potential Distribution

As presented in Figure 3.17, the apparent zeta potentials (ζ-potentials) of GO and three silica powder samples, measured in deionized water (pH = 7) at $T = 25^{\circ}C$, are displayed. All four particle types carried negative surface charges, which aligns with established understanding. For the silica powders, the magnitude of the zeta potential (negative) increased with larger particle size (e.g., D_{50} = 24.7 μm). Consequently, larger particles exhibit a greater propensity for attachment, as they are subject to enhanced electrostatic double-layer compression. Notably, the apparent zeta potential distribution for the silica powders with D_{50} = 0.8692 μm spans a broad range. The particles within this powder vary in size and approach the conventional 1 μm boundary, distinguishing colloidal particles from SPs. As a result, their attachment mechanism (i.e., the attachment characteristics) was influenced by the combined effects of both thermodynamic and hydrodynamic processes.

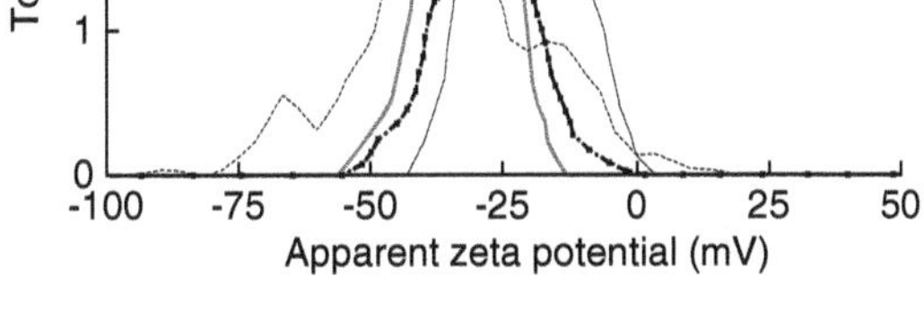

FIGURE 3.17 Apparent zeta potentials of GO and three silica powders.

Conversely, GO exhibited a narrower zeta potential distribution range. Furthermore, its mean zeta potential (D_{50} = 0.7287 μm; Figure 3.17) was lower than that of silica powders with a slightly larger particle size (i.e., D_{50} = 0.8692 μm). Consequently, GO demonstrated higher stability, which translated into a greater attachment affinity for surfaces. The observed disparities in the average potentials between GO and the silica powders primarily stem from differences in their apparent surface characteristics (Figure 3.10). Naturally, the strength of attachment is also significantly influenced by the structural properties of the particulate material.

3.4 REMEDIATION OF HMs IN WATER BY RMs

3.4.1 Materials and Schemes

3.4.1.1 Selected Raw RMs

The utilized RMs was sourced from Shandong, China, originating as a by-product from alumina refining via a combined process (Bai et al., 2022a). Following air-drying, the particles were oven-dried at 105 °C for 2 days. After being cooled to a set temperature of 22°C, the material was subsequently crushed and promptly sieved through a 0.25-mm sieve. The sieved samples were then stored in sealed plastic bags for subsequent analysis. The chemical composition of the RMs was analyzed using energy-dispersive X-ray (EDX) spectroscopy, revealing the following weight percentages: O in RMs accounted for 53.3% in weight, C 21.2%, Fe 7.4%, Na 6.5%, Al 5.6%, Si 3.9%, Ti 1.5%, and Ca 0.6%. X-ray diffractometry was used to measure the mineral composition of RMs. Quartz sand accounted for 10% by weight, diaspore 18%, chantalite 17%, boehmite 15%, hematite 13%, goethite 10%, calcite 11%, and lepidocrocite 6%.

FIGURE 3.18 SEM images. (a) RMs. (b) With Pb^{2+}. (c) With Cd^{2+}. (d) With Cu^{2+}.

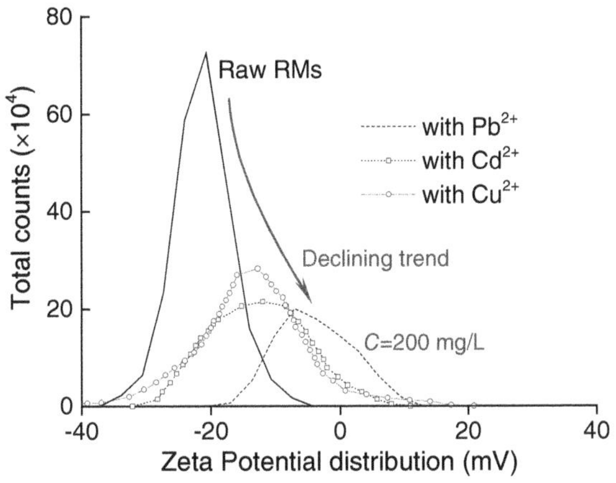

FIGURE 3.19 Zeta potential distribution of RMs loaded with HMs.

Particle size analysis indicated that 95% of the RMs fell within the range of 0.24–12.96 μm, while 60% were smaller than 3.11 μm. The SEM images of the RMs, presented in Figure 3.18a, reveal a microstructural framework composed of a cementitious matrix. This structure is predominantly formed through two binding mechanisms: cementation and crystallization. The particles exhibited considerable size heterogeneity, with a broad size distribution and surfaces characterized by roughness, porosity, and numerous protrusions and depressions. A larger specific surface area generally correlates with enhanced attachment-capacity. For the selected RMs, a specific surface area of 22.931 m^2/g was measured via a nitrogen adsorption porosimeter. The surface charge properties were investigated using a Malvern zeta potential analyzer, with the particle concentration maintained at 2 g/L. From Figure 3.19, the zeta potential of the RMs was measured at −21.2 mV (with a deviation of 4.48 mV). The particle charge distribution was relatively narrow, displaying a pronounced and high-intensity peak.

The surface of the RMs presented an overall negative charge. Moreover, the raw RMs were primarily found in a monodispersed state (Figure 3.18a). Consequently, these particles demonstrated substantial ion-exchange capability when interacting with positively charged HMs. Following the completion of the adsorption tests (i.e., loaded with HMs), the surface functional groups of the RMs were analyzed using FT-IR spectroscopy. This analysis was conducted to elucidate the adsorption mechanisms between the adsorbent (i.e., RMs) and the adsorbate (e.g., HMs).

3.4.1.2 HMs and Reagents

The Pb^{2+} solution was prepared by diluting a standard Pb^{2+} stock solution. Its primary constituent was $Pb(NO_3)_2$. This solution remained stable at ambient temperature, was odorless, and appeared as a colorless, transparent liquid with a pH ranging from 3.0 to 4.0. The Cu^{2+} solution was formulated using $(Cu(NO_3)_2)$, which was highly water soluble and exhibited relatively low toxicity. Similarly, the Cd^{2+} solution was synthesized from $(Cd(NO_3)_2 \cdot 4H_2O)$.

For the chromogenic detection of Cu^{2+}, sodium diethyldithiocarbamate was employed as the complexing agent. This reagent, with the chemical formula $(C_2H_5)_2$ NCSSNa·$3H_2O$, is characterized as a white, flaky crystal that is readily soluble in water and exhibits strong hygroscopicity. It reacts with Cu^{2+} in solution to form a stable yellow-brown colored colloidal complex. This complex remains stable for approximately 30 min, allowing for the rapid spectrophotometric determination of Cu^{2+} concentration.

3.4.1.3 Test Schemes

Case 1: The adsorption of single HMs by RMs. The effects of HMs concentration, RMs dosage, contact time, temperature, and pH on the attachment-capacity of RMs for three HMs (i.e., Pb^{2+}, Cd^{2+}, and Cu^{2+}) were investigated (Figure 3.20). Stock solutions of each HMs were prepared at a concentration of 1 g/L using deionized water and stored for later use. As required for each test, these stock solutions were diluted to the target experimental concentration (C). In a typical adsorption experiment, a specified amount of RMs (C_{red}) was introduced as the adsorbent into 100 ml of the HMs solution at a known initial concentration in an Erlenmeyer flask. After reaching adsorption equilibrium, the residual concentration (C_e) of the HMs was measured. Following adsorption in all tests, the mixture was centrifuged to separate the

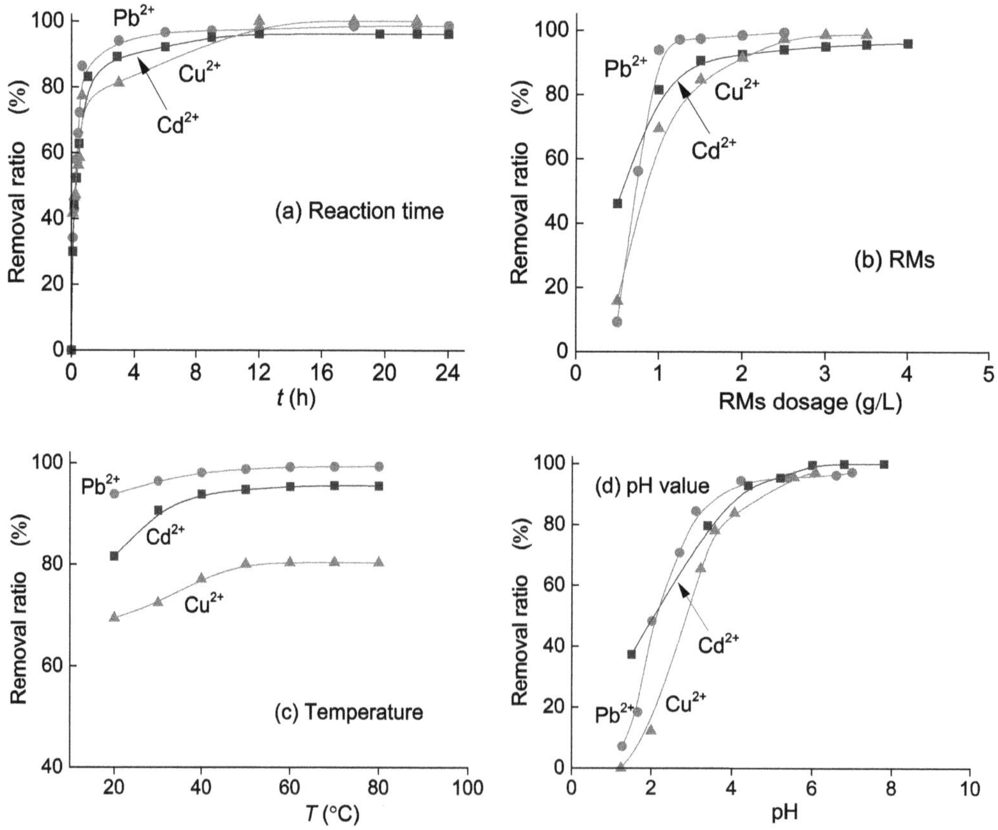

FIGURE 3.20 Influence of HMs on RMs. (a) Reaction time. (b) RMs dosage. (c) Temperature. (d) pH.

RMs from the solution. Three aliquots of the supernatant were then collected for analysis, and the average of these measurements was recorded as the final result.

Case 2: The competitive adsorption among different HMs (Cu^{2+}, Cd^{2+}, *and* Pb^{2+}). This series of experiments aimed to determine the adsorption preference order of RMs for different HMs and to examine the effect of adsorbent dosage on competitive adsorption behavior. The studies were conducted using RMs dosages of $C_{red} = 1$ and 2 g/L, initially HMs concentrations ranging from $C =$ 200 mg/L to 600 mg/L, a constant reaction time $t = 24$ h, and a temperature of $T = 20$ °C to assess the influence of RMs dosage. Each experimental condition was performed in duplicate, and results were averaged. For analysis, a total of six parallel supernatant samples were collected per test, and their average measurement was used to determine the outcome.

Case 3: The static attachment–detachment experiment for different HMs. This series of experiments examined the influence of HMs type (Cu^{2+}, Cd^{2+}, and Pb^{2+}) and temperature ($T = 20$ °C, 40 °C, and 60 °C) on the attachment and subsequent detachment behavior. The attachment phase employed a batch method, wherein 0.1 g of RMs was combined with 100 ml of HMs solution. After reaching attachment equilibrium over 24 h, the residual HMs concentration was measured. Following adsorption, a sequential detachment procedure was initiated. Also, 25 ml of the post-adsorption mixture was centrifuged. The supernatant was filtered to remove dissolved HMs, and the collected solid particles (i.e., RMs) were transferred to a new Erlenmeyer flask. Then, 25 ml of deionized water was added to initiate detachment. This process was repeated sequentially. By the seventh detachment cycle, all RMs had been isolated via filtration. These particles were then introduced into a final volume of 100 ml of deionized water for one last detachment step.

3.4.1.4 Test Apparatus and Procedure

During the experiment, an oscillation frequency of 100 r/min was maintained. Separation of HMs from RMs was achieved by centrifugation, utilizing a centrifuge with a maximum speed of 6500 rpm. For this procedure, the operational speed was set to 4500 rpm. A constant-temperature water bath equipped with a stirring function provided precise temperature control, with an accuracy of 0.5 °C.

For the determination of Pb^{2+} and Cd^{2+}, the measurements of their absorbance were conducted using a graphite furnace atomic-absorption spectrophotometer. Subsequently, based on the established correlation between absorbance and concentration, the absorbance readings were transformed into Pb^{2+} concentration. For the case of Cu^{2+}, the absorbance of its metal complex was measured with an ultraviolet (UV) spectrophotometer. These values were then converted into Cu^{2+} concentration by applying the corresponding absorbance–concentration relationship.

Adsorption-isotherm curves derived from static attachment tests are frequently utilized to evaluate the attachment characteristics of adsorbents toward HMs. The batch equilibrium method serves as a common approach for conducting static attachment experiments. The experimental procedure was carried out as follows. First, a solution containing HMs at a specified concentration was prepared. A measured quantity of raw RMs was then placed into an Erlenmeyer flask, followed by the addition of a predefined volume of HMs solution.

Subsequently, the flask was sealed and agitated in a mechanical shaker until the attachment process reached a state of equilibrium. After shaking, the mixture was transferred to a centrifuge tube and subjected to centrifugation to achieve complete solid–liquid separation. The resulting supernatant was carefully collected, and the equilibrium concentration of HMs present in it was determined.

3.4.2 Results and Discussion

3.4.2.1 Static Attachment of HMs

The attachment stability time serves as a key parameter for evaluating the adsorption performance of a given adsorbent. To quantify the removal efficiency of the adsorbent, the percentage removal is defined and utilized:

$$\eta = \frac{C - C_e}{C} \times 100\% \tag{3.17}$$

where η denotes the percentage removal, C denotes the initial HMs concentration, and C_e denotes the HMs concentration after reaching the adsorption equilibrium.

From Figure 3.20a, the rate at which the removal quantity increased for individual Pb^{2+}, Cd^{2+}, or Cu^{2+} ($C_{red} = 1$ g/L; C = 100 mg/L) showed a gradual deceleration over time. Once the contact period surpassed 1 h, the removal ratio swiftly approached a stable level; beyond 12 h, it remained largely constant. Upon attaining the adsorption equilibrium, the removal efficiencies reached 98.5%, 98.8%, and 95.1% of Pb^{2+}, Cd^{2+}, and Cu^{2+}, respectively. These results demonstrate that RMs possess a considerable capacity for the removal of HMs.

Experiment results revealed a marked increase in the removal of HMs as the dosage of RMs (i.e., the RMs concentration C_{red}; Figure 3.20b, C = 100 mg/L). This is attributed to the concomitant increase in available adsorption sites on the surfaces of RMs, leading to a rapid enhancement in adsorption efficiency. However, once the RMs dosage exceeded a certain threshold (e.g., $C_{red} = 1$ g/L), the rate of removal began to slow. This occurred because the number of adsorption sites provided by RMs surpassed the critical amount required for essentially complete adsorption of HMs, causing the removal rate to stabilize. The highest removal efficiency for Pb^{2+}, reaching 99.3%, was achieved at a RMs dosage of $C_{red} = 2.5$ g/L. For Cd^{2+}, the maximum removal of 96.0% was observed at $C_{red} = 4.0$ g/L. With increasing concentrations of HMs, the adsorption rate on RMs initially rose, but the final attachment quantity gradually approached saturation. These findings suggest that the adsorption of HMs onto RMs primarily proceeds via surface adsorption, mainly ion exchange, which may be interrelated with the strong alkalinity of RMs itself.

The effect of temperature on the removal efficiency of HMs by RMs is evident from Figure 3.20c ($C_{red} = 1$ g/L, C = 200 mg/L, T = 20–80 °C). Elevated temperatures were found to enhance the adsorption efficiency for all three HMs (i.e., Pb^{2+}, Cd^{2+}, and Cu^{2+}). For instance, upon reaching adsorption equilibrium, the removal percentages for Pb^{2+} at T = 20 °C, 40 °C, and 60 °C were 93.9%, 97.68%, and 99.3%, respectively. This trend is primarily due to the increased probability of HMs colliding with the solid RMs surfaces at higher temperatures, which correspondingly accelerates the reaction rate. Furthermore, since the removal rates for Cd^{2+} and Cu^{2+} were comparatively lower at reduced temperatures (e.g., T = 20°C), a rise in temperature appeared to produce a more pronounced improvement in their removal. Subsequently, the removed percentage gradually stabilized.

The presence of precipitation in the ionic solution further contributes favorably to HMs removal through mechanisms of coprecipitation and adsorption. Specifically, for Pb^{2+}, based on the precipitation equilibrium constant of $Pb(OH)_2$ (i.e., $K_{sp}=1.42\times10^{-20}$ at 25°C), the derived pH for the onset of Pb^{2+} precipitates is approximately 9.7. The surface electrical properties of RMs are closely linked to the solution pH, meaning pH also influences the speciation of Pb^{2+}. As depicted in Figure 3.20d, which examines the effect of initial pH on removal efficiency (in the range of 1–8), a pronounced increase in removal rate was observed with rising alkalinity when $1 < pH < 4$. For instance, at pH 4.3, the removal efficiency reached 94.5%. Under relatively low pH conditions, the highly acidic environment compromises the crystal structure of RMs, leading to a significant reduction in removal performance.

Regarding Cd^{2+}, with a K_{sp} for $Cd(OH)_2$ of 5.27×10^{-15}, the calculated precipitation pH is approximately 9.36. From Figure 3.20d, the removal percentage rose sharply from 37.4% to 92.8% as the pH increased from 1.2 to 4.4, after which it gradually stabilized. The attachment quantity of RMs for Cd^{2+} was more effectively utilized when the pH level exceeded 5.0. Similarly, for Cu^{2+}, the K_{sp} of $Cu(OH)_2$ is 4.37×10^{-20}, which corresponds to a precipitation pH of about 6.82. When the pH increased from 1.2 to 3.6, the removal percentage showed a sharp rise from nearly 0 to 78.1% (Figure 3.20d) before reaching a plateau. The removal rate of Cu^{2+} by RMs was maximized when the pH exceeded 6.0.

Based on the trends observed in Figure 3.20b and d, a further increase in RMs dosage and higher alkalinity enhanced the adsorption efficacy of RMs toward Pb^{2+} and Cd^{2+}, whereas Cu^{2+} adsorption achieved a stable state earlier. In Figure 3.20d, the pH values correspond to the initial measured pH of the prepared solution. It is noteworthy that the pH also shifted after adsorption concluded. In practice, the adsorption rate of HMs by RMs surpassed 90% when the solution pH was within the neutral range. Consequently, the application of RMs for adsorbing HMs relies on their intrinsic ion-exchange capability rather than requiring highly alkaline conditions. The mineral composition and structural characteristics of the soil matrix (e.g., hematite, goethite) contribute positively to adsorption efficiency, as does the presence of functional groups in porous materials. Here, the zeta potential of RMs surfaces was measured after loading with HMs. The HMs concentration was maintained at 200 mg/L, and the RMs concentration was set at 2 g/L. As illustrated in Figure 3.19, following the loading of HMs (Pb^{2+}, Cd^{2+}, and Cu^{2+}), the peak of the zeta potential distribution decreased markedly, and the distribution range broadened. Overall, the surface charge of the RMs shifted toward positive values (i.e., the distribution curve shifted to the right), suggesting that different particles tended to aggregate, forming clusters of varying sizes. This indicates that the charge properties on the RMs surfaces resulted in a wider distribution of both positive and negative charges. These findings demonstrate that after loading with HMs, the assembly of RMs transitions from an initial state characterized by singular cementation/crystalline connections (Figure 3.18a) to a turbid, flocculation-like structural form (Figure 3.18b–d). For instance, Pb^{2+} -loaded RMs exhibited a surface zeta potential of −13.9 mV (zeta deviation = 5.73 mV, Figure 3.19) with a substantial proportion of the particle surfaces having acquired a positive charge. Naturally, under these conditions, the capacity of the RMs surfaces to adsorb additional HMs progressively diminished.

3.4.2.2 Influence of two HMs on Competitive Adsorption

Figure 3.21 illustrates the effect of increasing RMs dosage on the purification capacity for HMs within three binary systems: Pb^{2+}-Cd^{2+}, Pb^{2+}-Cu^{2+}, and $Cd^{2+}-Cu^{2+}$ ($C = 200$ and

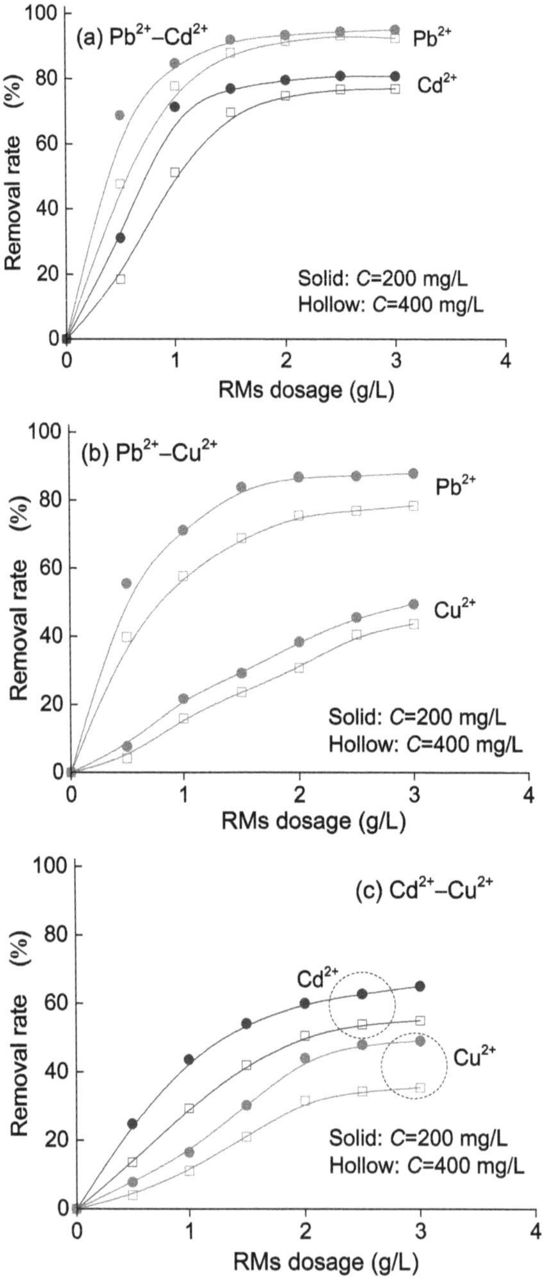

FIGURE 3.21 Competitive adsorption on removal ratio. (a) Pb^{2+}–Cd^{2+}. (b) Pb^{2+}–Cu^{2+}. (c) Cd^{2+}–Cu^{2+}.

400 mg/L). In the $Pb^{2+}-Cd^{2+}$ system, as the RMs dosage rose from C_{red} = 0-3 g/L, the removal efficiencies for both HMs increased gradually, with Pb^{2+} consistently exhibiting higher removal than Cd^{2+} (Figure 3.21a). Beyond a dosage of 1.0 g/L, the influence of further RMs addition gradually stabilized. Additionally, removal efficiencies declined somewhat when the total HMs concentration increased from 200 mg/L to 400 mg/L. For instance, at 200 mg/L, the maximum removal percentages reached 95.0% for Pb^{2+} and 80.8% for Cd^{2+}, whereas at 400 mg/L, the corresponding values were 92.4% and 76.9%. These results confirm that RMs exhibit a stronger adsorption affinity for Pb^{2+} compared to Cd^{2+} in mixed-ion conditions.

In the $Pb^{2+}-Cu^{2+}$ binary system, the removal efficiencies for both HMs rose progressively as the RMs dosage was elevated (Figure 3.21b), with Pb^{2+} consistently demonstrating a significantly higher adsorption rate than Cu^{2+}. For instance, at a total mixed solution concentration of 200 mg/L, the maximum removal percentages reached 87.9% for Pb^{2+} and 49.5% for Cu^{2+}. When the total HMs concentration was increased to 400 mg/L, the corresponding maximum removal values declined to 78.4% and 43.7%. Throughout this range, the removal rate continued to show an increase with higher RMs additions. Similarly, in the $Cd^{2+}-Cu^{2+}$ mixed solution (Figure 3.21c), the removal percentages of both HMs gradually improved with increasing RMs dosage, and the removal performance for Cd^{2+} remained superior to that for Cu^{2+}.

The comparative trends depicted in Figure 3.21b and c reveal that the removal efficiency in the $Cd^{2+}-Cu^{2+}$ binary system was lower than that in the $Pb^{2+}-Cu^{2+}$ system, which in turn was less effective than the binary system of Pb^{2+} and Cd^{2+}. When the concentrations in the $Cd^{2+}-Cu^{2+}$ mixed solution were 200 mg/L, the maximum removal percentages achieved were 65.1% for Cd^{2+} and 49.1% for Cu^{2+}. This behavior can be attributed to the fact that when two HMs coexisted, the ion for which RMs has a weaker adsorption affinity (e.g., Cu^{2+}) increases the overall activity of the mixed solution. As a result, the attachment quantity of RMs for the other HMs (e.g., Pb^{2+}) is correspondingly reduced.

3.4.2.3 Competitive Adsorption of the Three HMs

Evidently, in the three-HMs mixed solutions containing Pb^{2+}, Cd^{2+}, and Cu^{2+} (i.e., the ternary system), the removal efficiencies rose with increasing RMs dosage (Figure 3.22). The Pb^{2+} exhibited the highest purification efficiency, while Cu^{2+} showed the lowest. Once the RMs dosage exceeded 2 g/L, the removal efficiencies for Pb^{2+} and Cd^{2+} progressively stabilized. At a total concentration of 200 mg/L, the maximum removal percentages reached 93.9% for Pb^{2+}, 83.4% for Cd^{2+}, and 50.1% for Cu^{2+}. When the concentration increased to 400 mg/L, the corresponding maximum removal values declined to 89.1%, 74.7%, and 37.0%, respectively. The overall removal rate in the ternary system was slower compared to that in the binary systems (Figure 3.21), which in turn was slower than in single ion systems (Figure 3.20b). The adsorption affinity followed the order $Pb^{2+} > Cd^{2+} > Cu^{2+}$, a sequence that aligns with their measured zeta potentials (Figure 3.19). This preferential adsorption order is also consistent with observations reported for other adsorbents, such as artichoke agrowaste biomass. The underlying mechanism relates to the ionic radius, hydrated ion

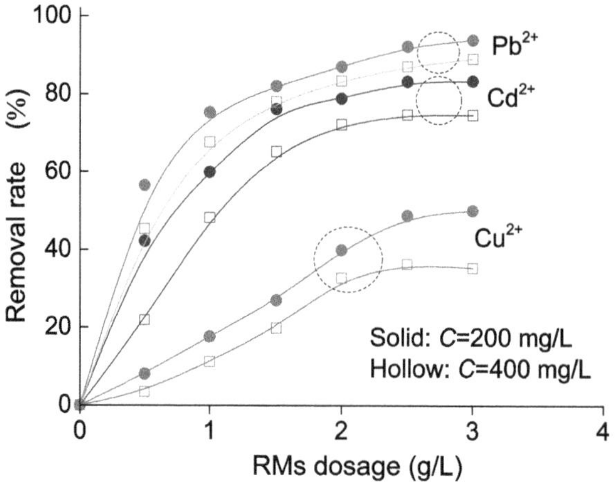

FIGURE 3.22 Competitive adsorption of $Pb^{2+}-Cd^{2+}-Cu^{2+}$ on removal ratios.

radius, and electron-sensing ability of each HMs. The Pb^{2+} possesses a relatively large ionic radius, which enhances its collision probability with active sites, while its large radius also corresponds to a weaker hydrated layer compared to Cd^{2+} and Cu^{2+}. Furthermore, charge characteristics are a key factor influencing adsorption affinity. Among the three HMs, Pb^{2+} has the highest electronegativity, leading to stronger electrostatic attraction to the surfaces of RMs.

3.4.2.4 Characteristics of Isothermal-Adsorption

The equilibrium adsorption is defined to characterize the adsorption ability of the adsorbent:

$$q_e = \frac{(C - C_e)V}{m} \tag{3.18}$$

where q_e denotes the equilibrium attachment quantity, V denotes the solution volume, and m signifies the adsorbent dosage.

Isothermal-adsorption experiments for a single HMs (i.e., Pb^{2+}, Cd^{2+}, or Cu^{2+}) onto RMs ($C_{red} = 2$ g/L) demonstrated that the attachment quantity at equilibrium (i.e., q_e, mg/g) rose with the equilibrium concentration of the ion (C_e, mg/L) and gradually approached a plateau (Figure 3.23a). Under binary competitive conditions between Pb^{2+} and Cd^{2+}, at RMs dosages of $C_{red} = 1$ and 2 g/L (Figure 3.23b), competitive adsorption of both HMs intensified with increasing ion concentrations before gradually tending toward saturation. In this system, RMs exhibited a stronger adsorption affinity for Pb^{2+} than for Cd^{2+}. For the binary competition involving Pb^{2+} and Cu^{2+} (Figure 3.23c), the attachment quantity for Cu^{2+} stabilized rapidly (i.e., the curve was flat), whereas the adsorption of Pb^{2+} increased considerably before gradually leveling off. Here as well, Pb^{2+} adsorption was notably superior to that of Cu^{2+}. In the $Cd^{2+}-Cu^{2+}$ binary system, the adsorption curves

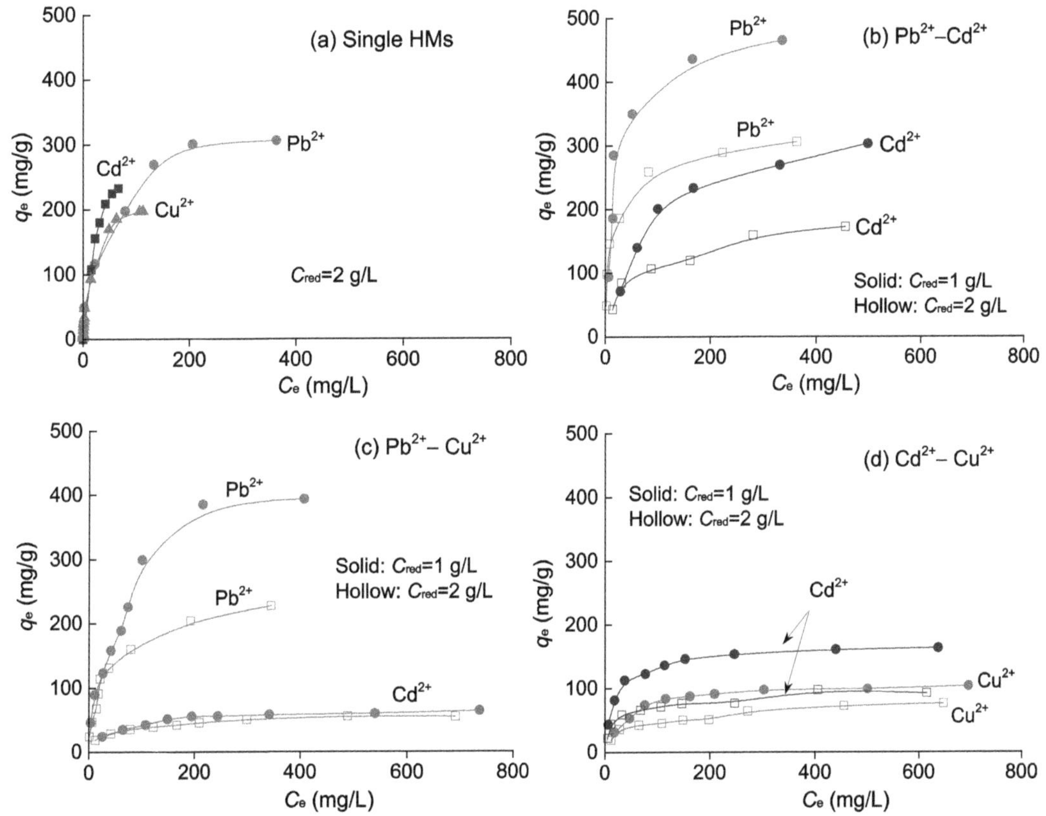

FIGURE 3.23 Isothermal-adsorption curves of RMs. (a) Single HMs. (b) $Pb^{2+} - Cd^{2+}$. (c) $Pb^{2+} - Cu^{2+}$. (d) $Cd^{2+} - Cu^{2+}$.

for both HMs stabilized quickly (Figure 3.23d), with RMs adsorbing Cd^{2+} more effectively than Cu^{2+} as the process gradually reached saturation. Overall, the trend observed in the isothermal-adsorption process aligns with the attachment-capacity patterns reported for many other adsorbents toward various ions. This consistency provides a useful basis for applying the theoretical description of the attachment–detachment process discussed subsequently.

From Figure 3.23, within a competitive system involving multiple HMs coexisting ions, the isothermal-adsorption curves for Pb^{2+}, Cd^{2+}, and Cu^{2+} were generally lower than those observed for a single HMs (Figure 3.23a). This effect was particularly pronounced when Pb^{2+} or Cd^{2+} coexisted with relatively weakly adsorbed Cu^{2+} (Figure 3.23c and d), aligning with the experimental trend noted for removal rates (Figure 3.21). Figure 3.24 presents the isothermal-adsorption curves for the ternary competitive system of $Pb^{2+} - Cd^{2+} - Cu^{2+}$ at RMs dosages of $C_{red} = 1$ and 2 g/L, where the adsorption capacities for all three HMs were reduced compared to any single-HMs system.

HMs are in a free state under acidic conditions (pH < 6). In the process of adsorption of Pb^{2+} by RMs, some substances (e.g., $CaCO_3$), possessing approximately 10% of the total

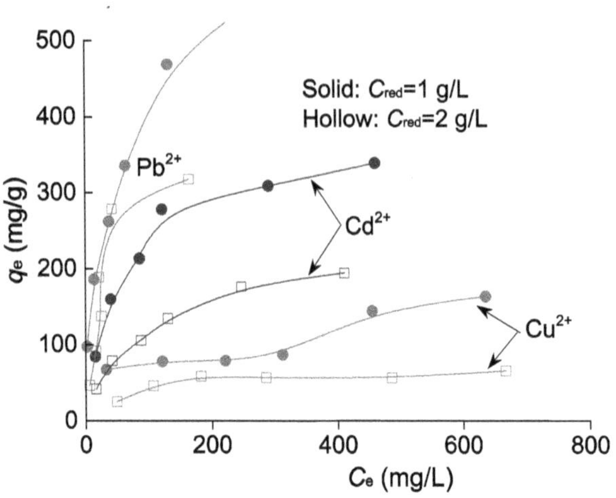

FIGURE 3.24 Isothermal-adsorption curves of Pb^{2+}, Cd^{2+}, and Cu^{2+} on RMs.

RMs in the mineral content of RMs, are completely dissolved at pH < 8, so there will be $PbCO_3$ precipitates generated, namely,

$$Pb(OH)^+ + CaCO_3 \rightarrow PbCO_3 + Ca^{2+} + OH^- \tag{3.19}$$

Under relatively high alkalinity (e.g., 8 < pH < 12), Pb^{2+} will also be converted into the form of its precipitate $Pb(OH)_2$. Consequently, Pb^{2+} adsorbed on RMs will also undergo the following reaction:

$$Pb(OH)^+ + OH^- \rightarrow Pb(OH)_2 \tag{3.20}$$

RMs is rich in chantalite, and its aluminosilicate minerals can interact with Cd^{2+} to form $CdSiO_3$. Furthermore, goethite (i.e., $FeO(OH)$), which possesses a substantial specific surface area and high surface reactivity, contributes significantly to this process. The mineral constituents of RMs, which remain stable under alkalinity, are generally effective at immobilizing HMs (e.g., Cd^{2+}). For example, during adsorption, goethite ($FeO(OH)$) may react with Cd^{2+} as follows to generate the complex $FeOOCd^+$, which subsequently precipitates:

$$FeO(OH) + Cd^{2+} \rightarrow FeOOCd^+ + H^+ \tag{3.21}$$

Furthermore, the amorphous hematite (Fe_2O_3) present in RMs will generate FeOH groups that impart high surface reactivity. As the pH rises, the variable negative charge on the Fe_2O_3 colloid increases, enhancing its electrostatic attraction. Consequently, its attachment-capacity for Cd^{2+} and Cu^{2+} is markedly enhanced. For instance, the surface FeOH groups on Fe_2O_3 can react with Cd^{2+} in solution as follows:

$$FeOH + Cd^{2+} \rightarrow FeOCd^{+} + H^{+} \tag{3.22}$$

The resulting complex ($FeOCd^{+}$) will precipitate on RMs surface to purify the water. The FeOH groups can also react with Cu^{2+} as follows:

$$FeOH + Cu^{2+} \rightarrow FeOCu^{+} + H^{+} \tag{3.23}$$

$$2FeOH + Cu^{2+} \rightarrow (FeO)_2 Cu + 2H^{+} \tag{3.24}$$

As a result, the surface coordination compounds formed (e.g., $FeOCu^{+}$, $(FeO)_2 Cu$) are also subject to precipitation. When RMs serve as an adsorbent material, the complex interactions among various ions reflect the multifaceted nature of its adsorption mechanism, often exhibiting synergistic effects. Additionally, the calcite ($CaCO_3$) present in RMs, which is rich in calcium carbonate, enables Cu^{2+} to form precipitates on RMs surface, leading to its effective enrichment.

RMs comprises varying mineral compositions and proportions, which correspondingly influence their adsorption (or complexation) efficacy toward different ions as well as the intensity of such effects. Consequently, the underlying adsorption mechanisms also differ. Moreover, the pH level of the RMs solution further modulates the mechanism by which HMs are adsorbed.

From another perspective, the introduction of deionized water into the solution (e.g., the HMs concentration decreases and the pH value decreases) causes some of the adsorbed substances (i.e., precipitates) to dissolve back into the aqueous phase as ionic species such as Pb^{2+}, Cd^{2+}, or Cu^{2+}. Consequently, a degree of detachment occurs under these conditions.

When compared to the SEM image of the original RMs (Figure 3.18a), the surfaces of RMs loaded with Pb^{2+} (as well as Cd^{2+} and Cu^{2+}) developed substantial amounts of hazy, floc-like deposits due to the electrostatic adsorption of HMs (Figure 3.18b–d). A comparable phenomenon has been observed in the SEM of various other adsorbents, including the adsorption of Cd^{2+}/Pb^{2+} onto modified argillaceous limestone, the adsorption of Cu^{2+}/Zn^{2+} by X zeolite, and the adsorption of Pb^{2+} (or Cd^{2+}) by an adsorbent zero-valent iron nanocomposite. Furthermore, iron-bearing minerals in RMs (e.g., lepidocrocite, hematite, and goethite) account for one-third of the total amount of RMs minerals, and iron-bearing minerals (e.g., goethite) contribute extensive surface area and pronounced surface activity, all of which significantly improve the overall adsorption ability of RMs.

3.4.2.5 Order of Adsorption Preference

The isothermal-adsorption curves obtained for the coexistence of the three HMs were different from those of a single HMs, a disparity primarily due to competition for adsorption sites and antagonistic interactions among the HMs. A partition coefficient k_p can be defined to quantitatively represent the preferential adsorption sequence:

$$k_p = \frac{q_e}{C_e} \tag{3.25}$$

TABLE 3.1 Partition Coefficients of HMs in the Presence of Other Competitive Ions

HMs	Concentration C (mg/L)	RMs, When $C_{red}=1$ g/L k_p (ml/mg)					RMs, When $C_{red}=2$ g/L k_p (ml/mg)				
		Binary, with			Ternary	$\sum k_p$	Binary, with			Ternary	$\sum k_p$
		Pb^{2+}	Cd^{2+}	Cu^{2+}			Pb^{2+}	Cd^{2+}	Cu^{2+}		
Pb^{2+}			13.71	3.81	13.48	31.00		36.08	5.14	5.61	46.83
Cd^{2+}	200	2.34		1.59	3.99	7.92	2.77		0.97	1.89	5.63
Cu^{2+}		0.35	0.72		0.64	1.71	0.32	0.41		0.43	1.16
Pb^{2+}			6.95	2.98	5.21	15.14		6.89	2.02	8.66	17.57
Cd^{2+}	400	1.41		0.62	2.28	4.31	0.74		0.31	1.03	2.08
Cu^{2+}		0.17	0.32		0.28	0.77	0.17	0.24		0.20	0.61
Pb^{2+}			2.66	1.79	3.58	8.03		1.30	0.66	1.94	3.90
Cd^{2+}	600	0.82		0.37	1.06	2.25	0.38		0.15	0.47	1.00
Cu^{2+}		0.11	0.20		0.32	0.63	0.08	0.12		0.10	0.30

where k_p (ml/mg) is the partition coefficient.

Experimental results indicated that under ternary adsorption conditions, the adsorption affinity of RMs followed the sequence $Pb^{2+} > Cd^{2+} > Cu^{2+}$. For example, at $C_{red} = 1g/L$ and a total concentration $C = 200$ mg/L, the corresponding coefficients were $k_d = 13.48$, 3.99, and 0.64 ml/mg, respectively (Table 3.1). Consequently, the overall attachment-capacity for each HMs can be comprehensively assessed based on the order of the summed distribution coefficients $\sum k_d$ across binary and ternary systems, which consistently yields the hierarchy $Pb^{2+} > Cd^{2+} > Cu^{2+}$. This preference is fundamentally linked to the mineralogy of RMs: minerals such as lepidocrocite, goethite, and hematite, which are relatively abundant, exhibit a strong affinity for Pb^{2+}, whereas minerals that preferentially adsorb Cu^{2+} are less prevalent, resulting in the weakest adsorption for Cu^{2+}.

When a kind of HMs with weak adsorption affinity coexisted in the system, the partition coefficient of the more strongly adsorbed HMs was reduced (Table 3.1). For example, when Cu^{2+} or Cd^{2+} (where the adsorption affinity $Cu^{2+} < Cd^{2+}$) coexisted with Pb^{2+} ($C_{red} = 1$ g/L, $C = 200$ mg/L), the corresponding partition coefficient of Pb^{2+} was 3.81 and 13.71 ml/mg, respectively. Conversely, the presence of more strongly adsorbed coexisting HMs resulted in a higher partition coefficient for the more slightly adsorbed ions. For example, when Pb^{2+} or Cd^{2+} ($Pb^{2+} > Cd^{2+}$) coexisted with Cu^{2+} ($C_{red} = 1g/L$, $C = 200$ mg/L), the corresponding partition coefficient of Cu^{2+} was 0.35 and 0.72 ml/mg, respectively. Analogous trends were noted in experiments using different RMs concentrations combined with HMs. This pattern reflects a coupled competitive effect among the adsorbing species.

A reduction coefficient of the proportion removed is introduced to compare with the adsorption effect on a single HMs:

$$k_\eta = \left(1 - \frac{\eta_2}{\eta_1}\right) \times 100\% \tag{3.26}$$

TABLE 3.2 Reduction Rate of the Removal Proportion in the Presence of Competitive Ions

RMs (g/L)	HMs	Concentration C (mg/L)	Binary, Reduction Rate (%) with			Ternary, Reduction Rate (%)
			Pb^{2+}	Cd^{2+}	Cu^{2+}	
	Pb^{2+}			9.77	25.23	20.10
1	Cd^{2+}	200	12.55		46.51	26.53
	Cu^{2+}		68.74	76.21		74.64
	Pb^{2+}			13.80	36.08	24.92
1	Cd^{2+}	400	32.37		61.22	36.30
	Cu^{2+}		60.71	72.44		72.09
	Pb^{2+}			3.84	20.61	12.35
2	Cd^{2+}	200	10.42		35.47	13.51
	Cu^{2+}		41.29	39.69		37.35
	Pb^{2+}			5.57	12.21	15.94
2	Cd^{2+}	400	11.76		44.29	25.68
	Cu^{2+}		52.20	45.12		50.21

where k_η denotes the reduction rate of the removal proportion, η_1 denotes the removal proportion, and η_2 denotes the removal proportion under binary/ternary competition.

Within the competitive ternary system containing Pb^{2+}-Cd^{2+}-Cu^{2+}, the removal rates for the three HMs decreased compared to those in single-HMs systems (Table 3.2). The reduction coefficients k_r of removal declined in the order $Cu^{2+} > Cd^{2+} > Pb^{2+}$, indicating that the removal of Cu^{2+} was the most impacted by competitive adsorption in the ternary environment. Furthermore, as the RMs dosage increased, the reduction coefficient k_r exhibited a slight decline, implying that higher dosages of RMs somewhat alleviated this reduction and improved removal performance under competitive conditions.

CHAPTER 4

Experiments on the Cotransport of Multi-Substances

4.1 MAIN CONTENT OVERVIEW

This chapter examines the coupled transport of suspended particles (SPs) and heavy metal ions (HMs; e.g., lead ions [Pb^{2+}] and cadmium ions [Cd^{2+}]) within a porous medium column under varying seepage velocity, the injection concentration, and the size of suspended substances. The analysis focuses on how transport parameters, including the longitudinal dispersivity and attachment coefficient (or deposition coefficient) of SPs as well as the recovery rate of both HMs and SPs, change with the injected HMs concentration.

To explore this issue, laboratory soil column experiments are conducted. These experiments discuss the transport behavior of micrometer-sized silicon powders (representing SPs) and Pb^{2+} (representing HMs), both individually and in combination, through saturated porous media. The tests incorporate sudden changes in seepage velocity and injection concentration during the seepage process to clarify the underlying physical mechanisms. Additionally, the influence of temperature on the cotransport of multi-substances is considered.

Long column experiments are performed to analyze the absorption/deposition behavior and transport distance of HMs when SPs are present. After the experiment concluded, measurements are taken of the absorbed/deposited HMs and SPs along the column. Furthermore, the particle size distribution (PSD) and microstructure images at various transport distances are obtained.

4.2 TRANSPORT OF HMs IN THE PRESENCE OF SILICON POWDERS

4.2.1 Experimental Approach

The SPs suspension (i.e., the suspension of silicon powders) with Pb^{2+} was injected using a 60-ml syringe. Seepage water with controlled acidity (e.g., pH = 5.5 or pH = 7) was supplied to the experimental system from a water tank (Figure 2.1 in Chapter 2).

DOI: 10.1201/9781003783749-4

Quartz sands served as the porous medium, with a median grain diameter (d_g) of 2.2 mm and a size range of 0.5–3.5 mm. The grain-size fractions were approximately 2.3% for 0.5–1 mm, 36.1% for 1–2 mm, and 61.6% for 2–3.5 mm particles. The solid matrix had a density of $\rho_s = 2.65 \pm 0.05$ g/cm^3, and the porosity was n = 41.7%. Four types of spherical SiO_2 particles, with a median diameter of $D_{50} = 2.8\,\mu m - 25.5\,\mu m$ $(D_{50}/d_g = 0.0013 - 0.0116)$ and a density of $\rho_s = 2.20\,g/cm^3$ were employed as the injected SPs. These particles showed a unimodal PSD, and the particle size was in the range of 1–80 μm.

The experiments involved injecting SPs suspension with varying particle concentrations ($C_{inj} = 1$, 2, 4, and 8 mg/ml) and Pb^{2+} $(C_P = 0 - 800\,\mu g/ml)$ under different seepage velocities (v = 0.087, 0.173, and 0.260 cm/s). The resulting Reynolds numbers fell within the range of 1.12–3.34, ensuring laminar flow conditions throughout. The "pulse-injection" refers to an instantaneous injection pattern (Bai et al., 2019c). In every experiment, the SPs suspension of $V_{inj} = 30\,ml$ was injected in a pulse-injection fashion over a period of $t_{inj} = 2\,s$. All experiments were carried out at a consistent room temperature, maintained between 22 °C and 24 °C.

4.2.2 Effect of Acidity on the Transport of SPs

The transport behavior of SPs suspension is highly complex, primarily due to the mixed injection of contaminants like Pb^{2+} with SPs and the environmental change of seepage water (e.g., the variation of pH and Pb^{2+} concentration). Since Pb^{2+} predominantly exists in acidic environments (e.g., $pH = 5.5$), examining how acidity influences the coupled transport of SPs and Pb^{2+} is essential. Two SPs with different pH values were considered. One suspension was neutral (pH = 7), and the other suspension was adjusted to $pH = 5.5$ by adding 0.01 mol/L of nitric acid (HNO_3) solution. The pH value of seepage water in the tank was the same as that of the injected suspension ($pH = 7$ or 5.5). The relationships between the relative concentration (C_R) and pore volume (V_P), the concepts of which can be seen in Chapter 2, for different injected concentrations (C_{inj}) of SPs are given in Figure 4.1, where v = 0.087 cm/s. The peak values of breakthrough curves (BTCs) elevate with higher seepage velocity. Regarding particle concentration, they initially rise but then stabilize as C_{inj} continues to increase.

A decrease in pH $(pH = 7 \rightarrow 5.5)$ leads to more significant adsorption or deposition, attributed to the higher concentration of positively charged H^+ introduced with HNO_3. Indeed, suspension pH strongly affects the surface charge characteristics of both SPs and the porous matrix, thereby influencing SPs deposition. Furthermore, increased acidity $(pH = 7 \rightarrow 5.5)$ reduces the electrostatic repulsion between SPs and the solid matrix, resulting in lower BTC peaks, particularly for larger SPs. For instance, when $D_{50} = 2.8, 10.9$, and $18.7\,\mu m$, $C_{inj} = 8$ mg/ml, and $v = 0.087$ cm/s (Figure 4.1b), the peak values are $C_R = 1.64, 0.51$, and 0.29 for $pH = 7$, and $C_R = 1.31, 0.29$, and 0.07, for $pH = 5.5$, respectively.

At a constant seepage velocity, the BTC peaks diminish while the corresponding V_P shows a slight increase as particle size grows ($D_{50} = 2.8\,\mu m \rightarrow 10.9\,\mu m \rightarrow 18.7\,\mu m$; see Figure 4.1). This occurs because the population of matrix pores larger than the SPs

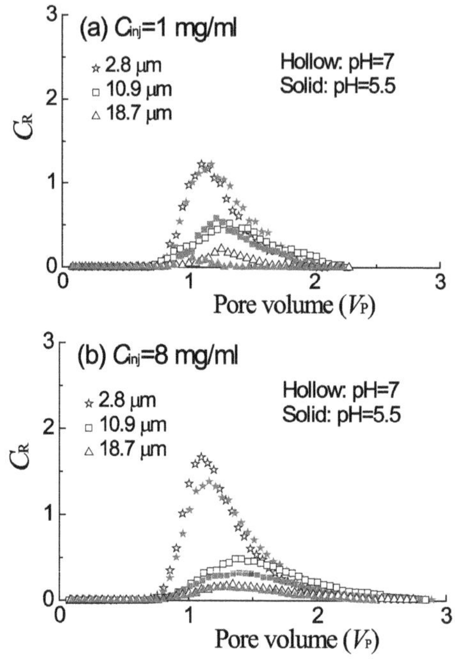

FIGURE 4.1 BTCs of SPs at v = 0.087 cm/s. (a) $C_{\text{inj}} = 1$ mg/ml. (b) $C_{\text{inj}} = 8$ mg/ml.

decreases with increasing the size of SPs, enhancing particle capture at pore throats and consequently reducing their effluent concentration peaks.

4.2.3 Seepage Transport of Lead Ions

The lead ions (Pb^{2+}) solutions of various concentrations were formulated using powdered $Pb(NO_3)_2$ and deionized water, with the pH adjusted to 5.5 using a 0.01 mol/L NaOH solution. To assess the potential influence of ionic strength (Na^+) caused by sample preparation, the transport of three injection concentrations of Pb^{2+} ($C_P = 100$, 300, and 500 μg/ml) was examined across three seepage velocities (v = 0.087, 0.173, and 0.260 cm/s). The seepage water pH was maintained at 5.5, identical to the injected solutions. The findings indicate that trace levels of Na^+ present a negligible impact on Pb^{2+} transport.

The penetration processes of Pb^{2+} under different concentrations and seepage velocities are presented in Figure 4.2. It reveals that the peak values of Pb^{2+} increase with increasing concentration of Pb^{2+} ($C_P = 100 \rightarrow 500$ μg/ml), but the corresponding V_P remains largely unchanged ($V_P = 1.04 - 1.08$). In contrast to SPs, the peak occurrence of Pb^{2+} in BTCs appears largely unaffected by seepage velocity. In fact, picometer-scale Pb^{2+} ions are completely soluble in water and move almost simultaneously with the water flow, which is mainly affected by the dispersivity and adsorption effect. The transport processes of SPs (Figure 4.1) are significantly affected by the so-called size exclusion effect, which makes the SPs plume have a lower effective dispersion coefficient than soluble Pb^{2+} due to the preferential advection of particles along the central streamline. Additionally, increased

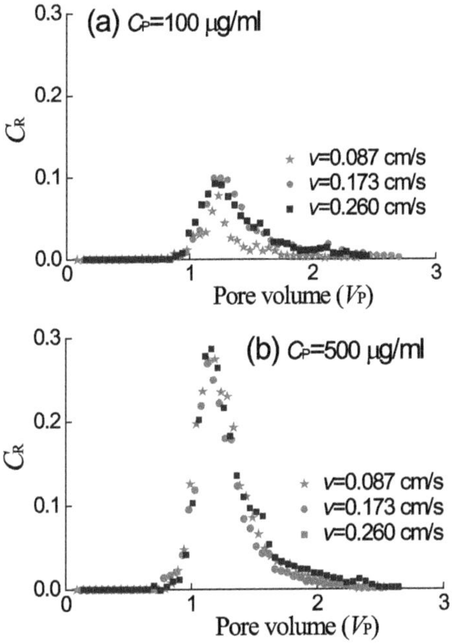

FIGURE 4.2 BTCs of Pb^{2+} for various seepage velocities. (a) $C_P = 100\,\mu g/ml$. (b) $C_P = 500\,\mu g/ml$.

seepage velocity enhances the hydrodynamic forces acting on SPs, reducing their deposition. Consequently, a greater number of SPs reach the outlet, leading to higher breakthrough peak values.

Figure 4.3 presents the transport parameters for Pb^{2+} as a function of injection concentration. The data indicate that the transport velocity of Pb^{2+} ($u/u_0 = 0.71-0.84$; Figure 4.3a) is consistently slower than the interstitial water velocity ($u/u_0 = 1$), and is also lower than the seepage velocity ($u/u_0 = 0.85-0.90$) reported for two natural SiO_2 particles ($D_{50} = 25$ and $47\,\mu m$) at comparable seepage velocities (v = 0.066–0.199 cm/s). In this context, u/u_0 represents the ratio of the interstitial velocity of Pb^{2+} (or SPs) to that of the water. This retardation can be explained by the tortuous transport paths of Pb^{2+} due to dispersivity, combined with significant adsorption onto the solid matrix.

The experimental results and the trends of dispersivity (α_d) and attachment coefficient (κ_d) across Pb^{2+} concentrations $C_P = 0-800\,\mu g/ml$ are also shown in Figure 4.3. A clear linear relationship exists between dispersivity α_d and Pb^{2+} concentration (Figure 4.3b; $R^2 = 0.77$), which is independent of seepage velocity. Increasing C_P leads to greater dispersivity, which in turn reduces the Pb^{2+} transport velocity to, for example, $u/u_0 = 0.72$ at a high injection concentration of $C_P = 800$ μg/ml (Figure 4.3a). In contrast, the attachment coefficients exhibit a slight decrease with increasing C_P (Figure 4.3c), but increase with higher seepage velocity.

This behavior is explained by the presence of positive charge (i.e., H^+ and Pb^{2+}) in the solution. According to the DLVO theory, these charges reduce the repulsive interactions

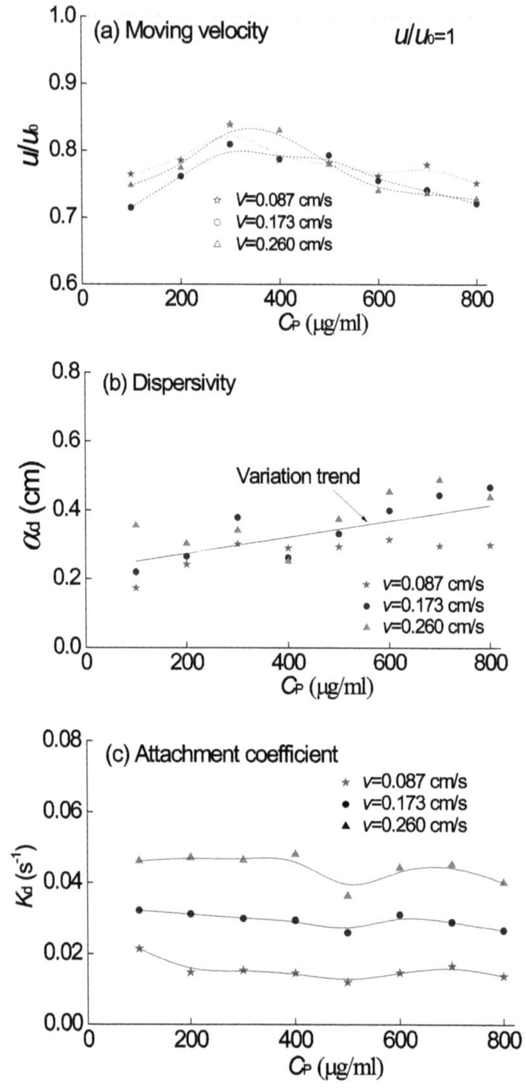

FIGURE 4.3 Transport parameters with injection concentration. (a) Moving velocity. (b) Dispersivity. (c) Attachment coefficient.

between Pb^{2+} and the matrix by decreasing the total interaction energy (i.e., the repulsive electrical double layer forces). Consequently, the combined effect of a lower velocity (u) and the increase of dispersivity (Figure 4.3b) promotes apparent adsorption (Figure 4.3c). The attachment coefficient increases by nearly a factor of 2, from $\kappa_d = 0.014 - 0.021\,s^{-1}$ to $\kappa_d = 0.036 - 0.048\,s^{-1}$, as the seepage velocity rises from v = 0.087 cm/s to v = 0.260 cm/s.

4.2.4 Coupled Transport of Lead Ions and SPs

The mixed suspension containing both Pb^{2+} and SPs was prepared according to the same methodology used for the Pb^{2+} solution. Figure 4.4 only illustrates the transport behavior

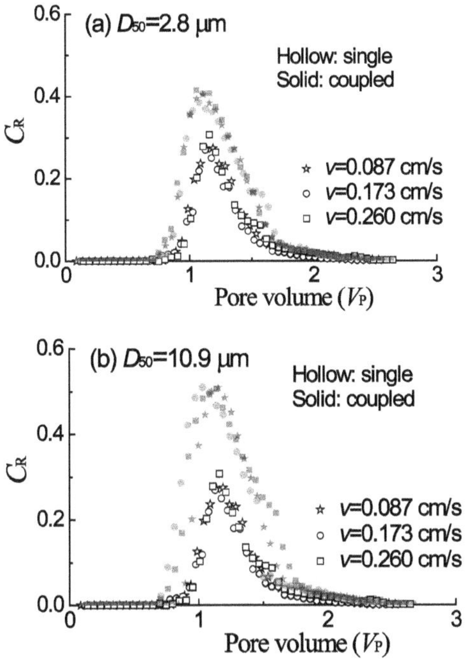

FIGURE 4.4 BTCs of Pb^{2+} during the coupled transport. (a) $D_{50} = 2.8\ \mu m$. (b) $D_{50} = 10.9\ \mu m$.

of Pb^{2+} under three seepage velocities (v = 0.087, 0.173, and 0.260 cm/s) and two sizes ($D_{50} = 2.8$ and $10.9\ \mu m$), comparing scenarios with SPs presence (called "coupled transport," solid marks) and absence (called "single transport," hollow marks) at pH = 5.5, $C_P = 500\ \mu g/mL$, and $C_{inj} = 2\ mg/ml$. As evidenced in Figure 4.4a, the presence of SPs markedly enhanced the BTCs peak values of Pb^{2+} and led to an earlier arrival (smaller V_P at peak). This accelerative influence stems from the role of SPs as carriers. Negatively charged surfaces of SPs possess a strong affinity for adsorbing positively charged HMs like Pb^{2+}. As these SPs are transported through the porous media at a rate comparable to, or exceeding, that in the aqueous phase, a phenomenon attributed to the size exclusion effect. The BTCs characteristics and peak values in this coupled system show minimal dependence on seepage velocity. Figure 4.4 indicates that the peak values are $C_R = 0.40$, 0.51, 0.44, and 0.36 in the presence of SPs for $D_{50} = 2.8$, 10.9, 18.7 and D_{50}=25.5 μm, respectively; but $C_R = 0.28$ on average in the absence of SPs. Here, the peak values are the mean values for the three velocities (v = 0.087, 0.173, and 0.260 cm/s).

The so-called promotion effect was previously observed in the coupled transport of colloids of <1 μm in size with various contaminants such as alkali and alkaline earth cations (Cs^{+} and Sr^{2+}), transition metals (Ni^{2+}, Co^{2+}, Cu^{2+} and Pb^{2+}), oxyanions (arsenic and iodate), and organic compounds. Whether SPs promote or retard contaminant transport depends on the complex interplay of coupled interactions, inherent SPs transport properties, hydrodynamic force, and suspension chemistry. SPs within an optimal size range (e.g., $D_{50} = 2.8 - 10.9\ \mu m$; Figure 4.4a and b) can substantially enhance Pb^{2+} transport via the size exclusion mechanism. However, SPs exceeding a critical size (e.g., $D_{50} = 18.7\ \mu m$, $D_{50}/d_g = 0.0085$) exhibit a diminished promotion effect or even inhibition of transport due to pore plugging, which reduces medium permeability. For instance, at $D_{50} = 25.5\ \mu m$ ($D_{50}/d_g = 0.0116$), the peak value of Pb^{2+} for $C_R = 0.36$ is only marginally higher than the $C_R = 0.28$ observed without SPs. This occurs because significant deposition of large SPs offsets the benefit of Pb^{2+} adsorption onto their surface. Pore plugging refers to the physical trapping of SPs at constrictive pore throats. The research suggests that this effect is primarily governed by a critical D_{50}/d_g threshold, typically within the range of $D_{50} / d_g = 0.0016 - 0.027$; beyond this threshold, SPs tend to retard contaminant transport. The severity of plugging is further influenced by SPs physical properties and PSD, matrix grain shape and roughness, pore-scale hydrodynamics, and solution chemistry.

Figure 4.5 presents the BTCs for the SPs of various sizes under both coupled (with Pb^{2+}) and single (without Pb^{2+}) transport conditions at pH = 5.5, $C_P = 500\ \mu g/ml$, and $C_{inj} = 2$ mg/ml. The data clearly indicate that the presence of Pb^{2+} reduces the peak breakthrough values for SPs. Further research indicates the recovery rate of SPs at the outlet decreases with the concentration of injected Pb^{2+} and gradually tends to be stable. The recovery rate is calculated by integrating simulated BTCs area, indicating the ratio of suspended substances in the effluent to the injected suspended substances. This effect is particularly pronounced at lower seepage velocities (e.g., v = 0.087 cm/s) and for larger sizes of SPs (e.g., $D_{50} = 18.7\ \mu m$). For example, with $D_{50} = 2.8\ \mu m$ (Figure 4.5a), the peak C_R values

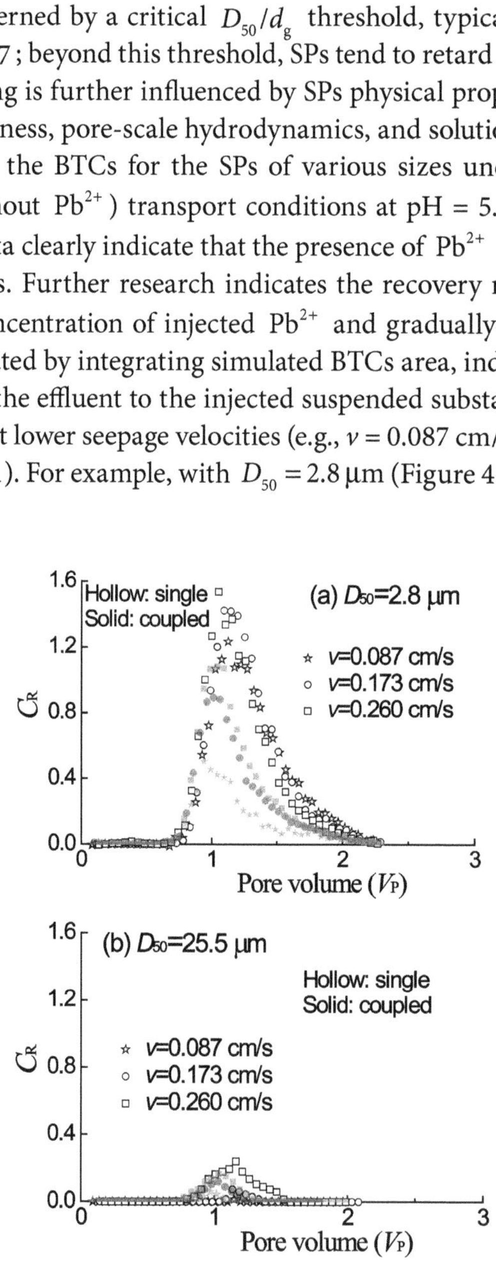

FIGURE 4.5 BTCs of SPs during the coupled transport. (a) D_{50} = 2.8 μm. (b) D_{50} = 25.5 μm.

in the presence of Pb^{2+} are 1.08, 0.89, and 0.51 for velocities of 0.087, 0.173, and 0.260 cm/s, respectively, compared to higher C_R values of 1.54, 1.42, and 1.23 in the absence of Pb^{2+}. This attenuation is primarily due to the adsorption of Pb^{2+} onto SPs surfaces, which reduces the repulsive electrostatic force between SPs and the matrix, thereby increasing their deposition potential. Consequently, the coupling effect of strongly charged ions like Pb^{2+} on SPs transport is significant and must be considered, stemming from a compression of the electrical double layer and a reduction in surface potential energy.

While the impact of SPs on contaminant seepage has garnered considerable interest, the reverse effect, how contaminants influence particle transport, has received less scrutiny. There is an increasing interest in the effects of SPs on the transport of HMs like Pb^{2+}. However, even trace amounts of specific contaminants can substantially alter particle transport dynamics by modifying surface charge properties, without significantly changing mass. In contrast to weakly polar or neutral contaminants, the coupling effect with strongly polar HMs like Pb^{2+} warrants particular attention for solid particle transport. Fundamentally, the influence of Pb^{2+} on SPs transport is governed by its ability to modify the physicochemical environment of the suspension, altering interfacial interactions and deposition behavior.

4.3 COTRANSPORT UNDER UNSTEADY FLOW AND VARIABLE INJECTION CONCENTRATIONS

4.3.1 Experimental Description

The coupled transport of relatively large SPs and soluble HMs was influenced by factors such as the hydrodynamics, surface charge, gravity, and particle size. Accounting for these factors, the SPs (i.e., a kind of silica powders) with median particle sizes of $D_{50} = 5.1$, 17.1, and 24.7 μm were chosen for these experiments. These particle sizes enable significant transport through the soil column while minimizing pore clogging, thereby maintaining relatively constant media porosity. To facilitate macroscopic observation during the experiments, light green, red, and black SPs with values of $D_{50} = 5.1$, 17.1, and 24.7 μm, respectively, were selected as SPs (Bai et al., 2021a).

Transport behaviors can differ substantially among various HMs. The Pb^{2+} were chosen as the representative HMs. To ensure high solubility, the Pb^{2+} solution was prepared using 1.0 mol/ml HNO_3, resulting in a colorless and odorless liquid. The required experimental concentrations were obtained by diluting a standard stock solution, yielding acidic solutions with a pH of 5.5.

The suspension was injected continuously. During injection, rapid changes in seepage velocity and injection concentration were introduced at predetermined times. Three sets of experiments were conducted. In the first set of experiments, the transport of SPs alone in saturated porous media was considered with a variable water seepage velocity or injection concentration of SPs. For convenience, the apparent water seepage velocity (i.e., Darcy velocity) was used to express the seepage velocity. This scheme involved two median particle sizes (D_{50}), four types of changes in the seepage velocity (v), two injection concentrations (C_S) of SPs, and two types of changes in the injection concentration of SPs. The selected parameters comprise possible situations that may be encountered in engineering. The second set studied the transport of Pb^{2+} alone, with variable seepage velocity or Pb^{2+} injection concentration (C_P), encompassing two patterns of seepage velocity change and two concentration levels. The third set investigated the coupled transport of SPs and Pb^{2+} by simultaneously altering seepage velocity and the Pb^{2+} concentration. This set considered three sizes of SPs, two patterns of seepage velocity change (at two seepage velocity levels), two concentrations of SPs, and two patterns of Pb^{2+} concentration variation.

4.3.2 Transport of SPs

Figure 4.6a demonstrates that the effluent concentration of SPs rises with increasing injected concentration ($C_S = 0.5 \rightarrow 2$ mg/ml). During an initial stage of constant flow (i.e., in Stage 1 of the penetrating process, $V_P = 0-2.75$), once steady-state seepage was established, the effluent concentration also increased at higher seepage velocity (v = 0.087, 0.173, 0.260 cm/s). Moreover, a sudden increase in seepage velocity at $V_P = 2.75$ triggered a rapid rise in the effluent concentration. When different change processes reached an identical final velocity (e.g., from v = 0.087 cm/s to v = 0.260 cm/s; from v = 0.173 cm/s to v = 0.260 cm/s), the resulting steady-state effluent concentrations were very similar. Conversely, a sudden decrease in seepage velocity at $V_P = 2.75$ caused a rapid decline in effluent concentration. The effluent values, when velocities dropped to the same lower level, were also comparable. For example, the value after the steady state was reached in Stage 1 with v = 0.087 cm/s (i.e., $V_P = 2.75$), and the value when the steady state was achieved with $v = 0.260$ cm/s $\rightarrow 0.087$ cm/s (i.e., approximately $V_P = 4.50$). Identical trends were observed for SPs with a D_{50} of 24.7 μm (Figure 4.6b). This behavior is explained by

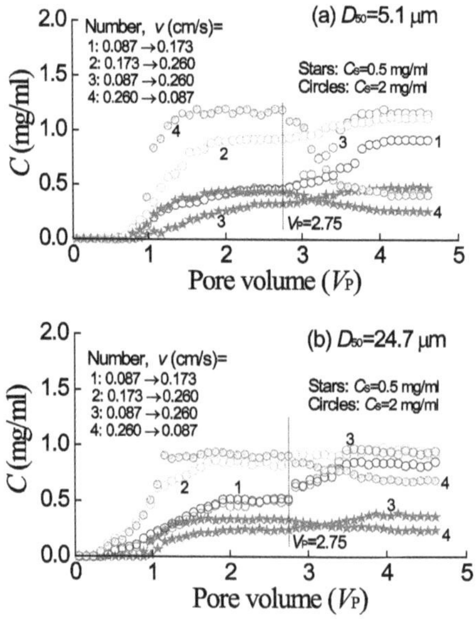

FIGURE 4.6 BTCs of SPs under unsteady flow. (a) $D_{50} = 5.1$ μm. (b) $D_{50} = 24.7$ μm.

hydrodynamic forces. Increased seepage velocity enhances the hydrodynamic effect on SPs, leading to a greater quantity of particles in the effluent. A seepage velocity reduction produces the opposite effect. The impact of changing seepage velocity on effluent concentration is due to a nonequilibrium hydrodynamic process governing SPs deposition and release, which controls mass transfer between deposited and mobile particles. This mechanism is further influenced by the specific physicochemical properties of both the SPs and the porous matrix.

Figure 4.6 demonstrates that the penetration process of SPs was significantly influenced by their particle size. For instance, at an injection concentration of $C_S = 2$ mg/ml and a relatively low seepage velocity (i.e., v = 0.087 cm/s), the effluent concentration of SPs with $D_{50} = 5.1$ and $24.7\,\mu\text{m}$ reached steady-state values of 0.45 mg/ml (Figure 4.6a, $V_P = 2.75$) and 0.49 mg/ml (Figure 4.6b), respectively, the former being slightly lower. In contrast, at a high seepage velocity (i.e., v = 0.260 cm/s), the steady-state concentration for SPs with $D_{50} = 5.1\ \mu\text{m}$ ($C = 1.15$ mg/ml, $V_P = 2.75$, Figure 4.6a) was substantially greater than that for $D_{50} = 24.7\ \mu\text{m}$ ($C = 0.90$ mg/ml, Figure 4.6b). These observations suggest that at low seepage velocity, the deposition of SPs was dominant, imposing considerable transport resistance. Under these conditions, smaller SPs experienced more extensive interfacial interaction with the matrix surface, facilitating their deposition. Meanwhile, larger SPs (e.g., $D_{50} = 24.7\ \mu\text{m}$) exhibited relatively higher penetration potential due to preferential flow paths, which could lead to a greater effluent concentration. Conversely, under the stronger hydrodynamic forces at higher seepage velocity, smaller particles could more readily traverse the smaller pores of the media, thereby increasing their number in the effluent. It is noteworthy that at a relatively low SPs injection concentration (e.g., $C_S = 0.5$ mg/ml), smaller particles ($D_{50} = 5.1\ \mu\text{m}$) consistently yielded a higher effluent concentration across all tested seepage velocities. This occurs because, at lower concentrations, irregular collisions between SPs and the matrix are reduced, which in turn enhances the overall transport process.

Figure 4.7a illustrates that for a given particle size of SPs (e.g., $D_{50} = 5.1$ or $24.7\ \mu\text{m}$), the penetration curves resulting from stepwise increases in injection concentration of SPs ($C_S = 0.1 \rightarrow 0.5 \rightarrow 2$ mg/ml) exhibited changes at pore volumes (V_P) of 2.75 and 4.61, respectively. The data indicate that once a steady-state concentration plateau was attained following each increase, the effluent concentration rose in accordance with the higher injected concentration. Notably, the rise in effluent concentration consistently lagged behind the step change in injection concentration (i.e., $V_P > 2.75$ or $V_P > 4.61$), a response pattern distinct from that induced by sudden seepage velocity (Figure 4.6). As expected, effluent concentration increased with higher seepage velocity, and this effect became more substantial at greater injection concentration. This suggests that intensified hydrodynamics conditions reduced the deposition potential of SPs, while gravitational settling became increasingly influential; collectively, hydrodynamics played a key role in facilitating SPs transport. For smaller SPs (e.g., $D_{50} = 5.1\ \mu\text{m}$), the effluent concentration increased at a decreasing rate as the injection concentration was raised. In contrast, for larger SPs (e.g., $D_{50} = 24.7\ \mu\text{m}$), the effluent concentration showed a marked and sustained increase with higher injection concentration.

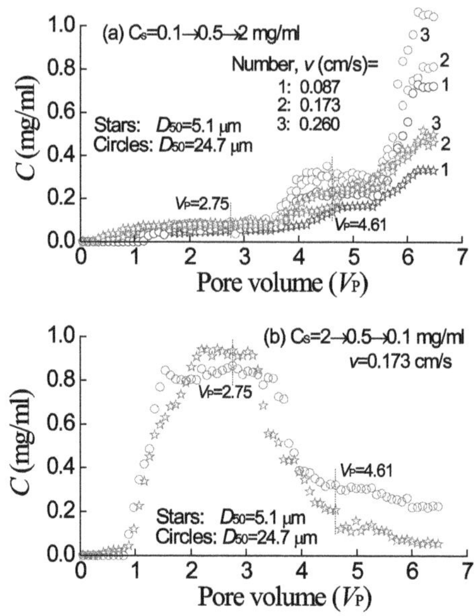

FIGURE 4.7 BTCs of SPs under variable injection concentrations. (a) $C_S = 0.1 \rightarrow 0.5 \rightarrow 2$ mg/ml. (b) $C_S = 2 \rightarrow 0.5 \rightarrow 0.1$ mg/ml.

When the injection concentration of SPs was progressively reduced (v = 0.173 cm/s, and $C_S = 2 \rightarrow 0.5 \rightarrow 0.1$ mg/ml; Figure 4.7b), a corresponding decline in effluent concentration was observed. At the initially high concentrations (e.g., $C_S = 2$ mg/ml, that is, Stage 1 of the penetration process, $V_P < 2.75$), large SPs (e.g., $D_{50} = 24.7$ μm) resulted in a lower effluent concentration compared to smaller ones (Figure 4.7b). However, after the concentration was lowered in a later stage (e.g., decreased to $C_S = 0.1$ mg/ml, that is, Stage 3 of the penetration process, $V_P > 4.61$), the effluent concentration for large SPs ($D_{50} = 24.7$ μm) exceeded that of smaller SPs ($D_{50} = 5.1$ μm). This shift occurs because the high particle concentration in Stage 1 promoted significant deposition/adsorption onto the solid matrix surface, an effect more pronounced for larger particles. Subsequently, reducing the injection concentration induced a detachment effect from the porous media, facilitating the removal of SPs that were clogging pore spaces—effectively a cleansing or purification process. The impact of changing injection concentration on effluent concentration is governed by a nonequilibrium hydrodynamic process, which is intrinsically linked to the attachment-capacity of the porous medium.

4.3.3 Transport of Lead Ions

Figure 4.8a presents the effluent concentrations of lead ions (Pb^{2+}) under different seepage velocities, including scenarios where the velocity was abruptly altered during the penetration process (e.g., from v = 0.087 cm/s to v = 0.260 cm/s; from v = 0.260 cm/s to v = 0.087 cm/s), with the pore volume V_P corresponding to a velocity change of 2.75. The results show that the steady-state effluent concentration and the pore volume required to

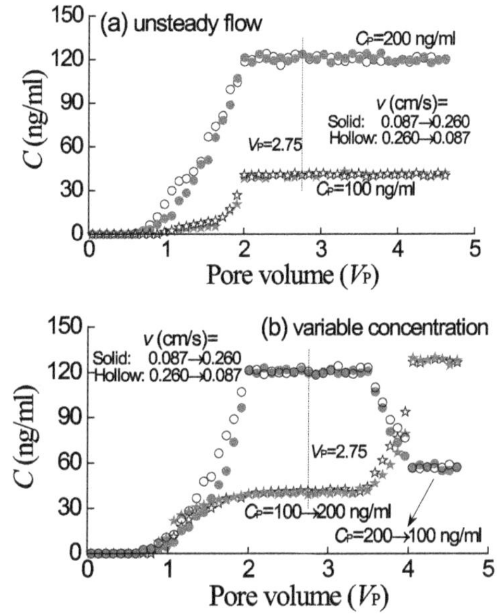

FIGURE 4.8 BTCs of Pb^{2+}. (a) Unsteady flow. (b) Variable concentrations.

reach it (i.e., $V_P = 2.1$) in Stage 1 remained virtually constant regardless of changes in seepage velocity. In other words, the transport of Pb^{2+} was insensitive to variations in flow rate. For example, at a Pb^{2+} injection concentration C_P of 100 ng/ml (Figure 4.8a), when the seepage velocity was increased from v = 0.087 cm/s to v = 0.260 cm/s (or decreased from v = 0.260 cm/s to v = 0.087 cm/s), the obtained steady-state concentration C was 40.5 ng/ml (or 40.6 ng/ml), and the pore volume V_P corresponding to the time when the effluent concentration reached the steady state was very clear (i.e., $V_P = 2.1$). This behavior occurs because Pb^{2+} is highly soluble in acidic solutions (e.g., solutions with a pH of 5.5), and its transport velocity through the porous media aligns with the seepage velocity of the water itself. Consequently, their breakthrough is independent of the specific pore volume V_P at which the flow rate change occurs.

From Figure 4.8b, under a constant seepage velocity (e.g., v = 0.260 cm/s), the effluent concentration of Pb^{2+} increased or decreased in response to a corresponding change in the injection concentration (e.g., $C_P = 100 \rightarrow 200$ ng/ml or $C_P = 200 \rightarrow 100$ ng/ml). Consistent with the earlier findings, for a given injection concentration, the steady-state effluent concentration and its corresponding pore volume V_P showed negligible variation with changes in seepage velocity. Furthermore, when the Pb^{2+} injection concentration was abruptly altered (e.g., at $V_P = 2.75$, Figure 4.8b), the temporal response of the effluent concentration (approximately $V_P = 3.50$) and the subsequent pore volume V_P for reaching the new steady-state (approximately $V_P = 4.00$) were largely independent of the seepage velocity. In this respect, the transport dynamics of dissolved Pb^{2+} are fundamentally distinct from those of larger, microscale SPs.

4.3.4 Coupled Transport of SPs and Lead Ions under Unsteady Flow

For relatively small SPs (e.g., $D_{50} = 5.1\ \mu m$; Figure 4.9), the presence of Pb^{2+} markedly altered their transport behavior, generally reducing their effluent concentration (C). This reduction persisted even when seepage velocity was abruptly changed (e.g., from $v = 0.087$ cm/s to $v = 0.260$ cm/s, or from $v = 0.260$ cm/s to $v = 0.087$ cm/s; the pore volume of seepage change was $V_P = 2.75$ cm) during the seepage penetration, mainly because the porous media matrix was made of quartz sand with a negatively charged surface. Upon mixing, the positively charged Pb^{2+} is adsorbed onto the negatively charged surfaces of SPs. This adsorption diminishes the surface charge polarity of the SPs, consequently weakening the electrostatic repulsion both among SPs and between SPs and the matrix. Smaller SPs possess a greater specific surface area, leading to more extensive adsorption of Pb^{2+}. Consequently, these more positively charged SPs are more likely to deposit onto the negatively charged matrix. This enhanced deposition results in a significantly lower effluent concentration for SPs in the presence of Pb^{2+}, compared to their transport alone. In contrast, for relatively large SPs (e.g., $D_{50} = 24.7\ \mu m$), Pb^{2+} exerts a diminished influence in transport. With larger particles, fewer Pb^{2+} adhere per unit surface area, and the movement of the SP/ Pb^{2+} mixture becomes increasingly governed by hydrodynamic processes rather than interfacial electrostatic interactions.

At relatively low injection concentrations of SPs (e.g., $C_S = 0.5$ mg/ml) and constant seepage velocity (in Stage 1 of the seepage process, $V_P < 2.75$), increasing the Pb^{2+} concentration

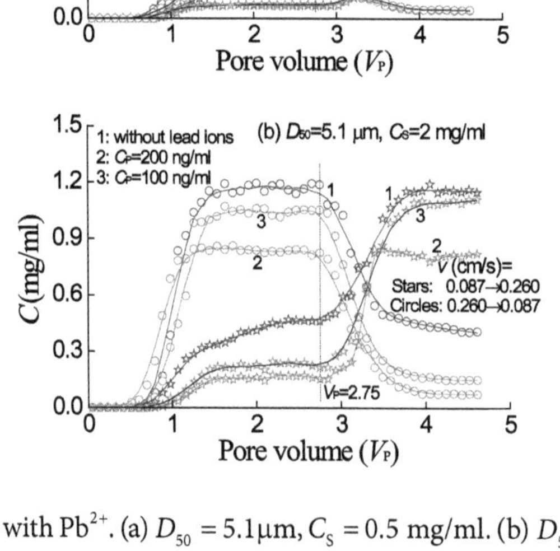

FIGURE 4.9 BTCs of SPs with Pb^{2+}. (a) $D_{50} = 5.1\mu m$, $C_S = 0.5$ mg/ml. (b) $D_{50} = 5.1\mu m$, $C_S = 2$ mg/ml.

(e.g., $C_P = 100 \rightarrow 200$ ng/ml) retarded the transport of SPs. This resulted in a more rapid attainment of steady-state and the correspondingly lower steady-state effluent concentration. Under higher concentrations of SPs (e.g., $C_S = 2$ mg/ml), the steady-state effluent concentration after the seepage stabilization decreased markedly with higher Pb^{2+} injection ($C_P = 100 \rightarrow 200$ ng/ml, that is, $V_P = 2.75$). Furthermore, this difference in steady-state concentrations became more pronounced at greater seepage velocity (in Stage 1 of the seepage process, $V_P < 2.75$). However, increased seepage velocity substantially enhanced the hydrodynamic driving forces acting on SPs.

The transport behavior of Pb^{2+} changed substantially when it was mixed with SPs, differing markedly from their transport alone. While the BTCs for Pb^{2+} alone remained unaffected by seepage velocity (Figure 4.10), the curve in the coupled system showed clear dependence on seepage velocity. At a relatively low Pb^{2+} concentration ($C_P = 100$ ng/ml), SPs significantly enhanced Pb^{2+} transport, with a more pronounced effect at higher seepage velocity. For example, at a concentration of $C_S = 0.5$ mg/ml (Figure 4.10a), the water flow changed from a low seepage velocity v = 0.087 cm/s to a high seepage velocity $v = 0.260$ cm/s in Stage 1 of the seepage process ($V_P < 2.75$), and the concentrations C of Pb^{2+} in the steady-state effluent corresponding to the cases without SPs and with SPs of $D_{50} = 5.1$, 17.1, and 24.7 μm were 40.5, 63.4, 46.4, and 40.9 ng/ml, respectively. This indicates a trend of decreasing Pb^{2+} concentration with increasing SPs size. At low SPs concentration and low velocity, smaller SPs adsorbed more Pb^{2+}, thereby facilitating greater contaminant removal. In Stage 2 of the experiment (i.e., the seepage velocity increased to $v = 0.260$ cm/s,

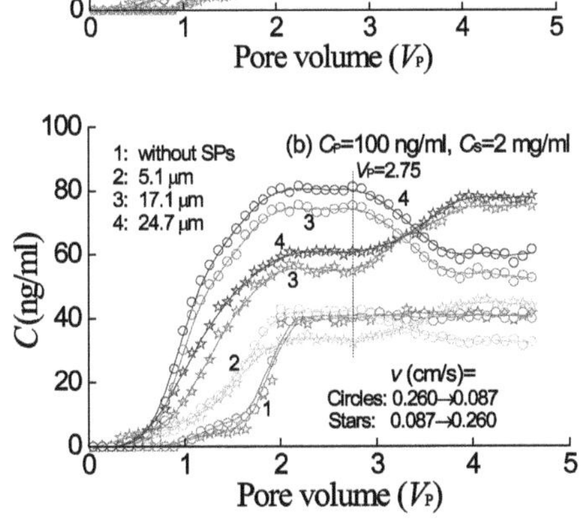

FIGURE 4.10 BTCs of Pb^{2+} under unsteady flow. (a) $C_P = 100$ng/ml, $C_S = 0.5$ mg/ml. (b) $C_P = 100$ng/ml, $C_S = 2$ mg/ml.

$V_P > 2.75$), the effluent concentration of Pb^{2+} corresponding to the cases without SPs and with SPs of $D_{50} = 5.1$, 17.1, and 24.7 μm were 40.5, 75.8, 64.6, and 50.8 ng/ml, respectively.

The Pb^{2+} in the effluent originated from two sources: those dissolved and transported with the aqueous phase, and those adsorbed onto SPs surfaces and carried by the mobile particles. Consequently, a sudden increase in seepage velocity from v = 0.087 cm/s to v = 0.260 cm/s ($V_P = 2.75$) enhanced the mobility of the SPs, which in turn increased the quantity of Pb^{2+} transported via this particulate pathway, thereby elevating the overall effluent concentration of Pb^{2+}. Conversely, a reduction in seepage velocity (from v = 0.260 cm/s to v = 0.087 cm/s; $V_P = 2.75$) led to a decrease in the effluent Pb^{2+} concentration. This trend appeared more noticeable at higher Pb^{2+} injection concentrations (Figure 4.10b).

However, at a relatively high Pb^{2+} concentration (e.g., $C_P = 200$ ng/ml), the presence of SPs tended to inhibit Pb^{2+} transport. In this scenario, the effluent concentration of Pb^{2+} in the absence of SPs was consistently higher than when SPs were present. Even when the seepage velocity increased from a low velocity v = 0.087 cm/s to a high velocity v = 0.260 cm/s, which raised the effluent concentration, the overall inhibitory effect of SPs on Pb^{2+} remained significant. Furthermore, this inhibitory influence became more pronounced with larger sizes of SPs (e.g., $D_{50} = 5.1 \rightarrow 17.1 \rightarrow 24.7$ μm), resulting in a further decrease in effluent concentration. In this case, a low concentration of SPs (e.g., $C_S = 0.5$ mg/ml) resulted in a relatively strong inhibitory effect, that is, a low effluent concentration of Pb^{2+}, whereas a high concentration of SPs ($C_S = 2$ mg/ml) carried more Pb^{2+} and therefore led to a higher concentration of Pb^{2+} in the effluent.

4.3.5 Coupled Transport of SPs and Lead Ions under Variable Concentrations

During the coupled transport of SPs and Pb^{2+}, elevating the injected Pb^{2+} concentration level resulted in reduced effluent concentration for small SPs (e.g., in Stage 1 of the seepage process, $V_P < 2.75$, $C_P = 200$ ng/ml, D_{50} = 5.1μm, marked by circles, Figure 4.11a). Moreover, the effluent concentration of SPs declined or rose following a sudden increase or decrease (e.g., $C_P = 100 \rightarrow 200$ ng/ml), respectively, in Pb^{2+} concentration during the experiment. For relatively large SPs (e.g., D_{50} = 24.7μm, Figure 4.11a), the influence of Pb^{2+} on the transport of SPs diminished. With two different Pb^{2+} concentrations (e.g., $C_P = 100$ and 200 ng/ml, Figure 4.11b) or a sudden change in the Pb^{2+} concentration during the transport process (e.g., $C_P = 100 \rightarrow 200$ ng/ml or $C_P = 200 \rightarrow 100$ ng/ml), the SPs penetration curves almost overlapped. At a relatively low SPs injection concentration ($C_S = 0.5$ mg/ml), the effect of the change in Pb^{2+} concentration on the transport of SPs was significantly reduced; in contrast, at a relatively high injection concentration (e.g., $C_S = 2$ mg/ml), especially under a relatively high seepage velocity such as v = 0.260 cm/s, the difference in effluent concentration became more pronounced. This is because SPs carrying adsorbed Pb^{2+} acquired a stronger positive charge, enhancing their attachment and detachment onto the negatively charged porous media surface.

Figure 4.12 examined the transport of Pb^{2+} due to an abrupt concentration change in the cases of $C_S = 0.5$ mg/ml and $C_S = 2$ mg/ml under the condition of v = 0.087 cm/s.

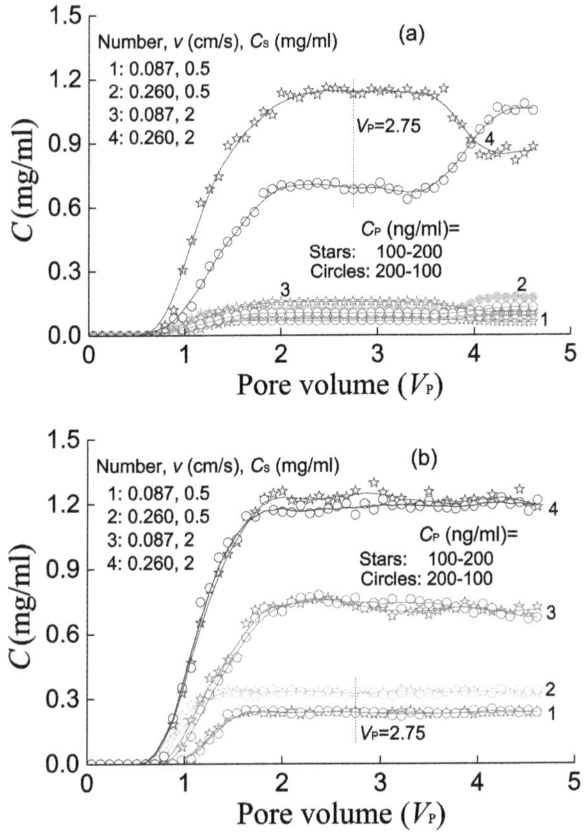

FIGURE 4.11 BTCs of SPs under different Pb^{2+} concentrations. (a) $D_{50} = 5.1$ μm. (b) $D_{50} = 24.7$ μm.

During the Pb^{2+} concentration increase ($C_P = 100 \rightarrow 200$ ng/ml), in Stage 1 ($V_P < 2.75$, $C_P = 100$ ng/ml), the effluent concentration of Pb^{2+} in the cases without SPs and with SPs of $D_{50} = 5.1$, 17.1, and 24.7 μm were 39.9, 63.4, 46.4, and 40.9 ng/ml, respectively. A clear trend of decreasing Pb^{2+} concentration with increasing SPs size was observed, indicating that the transport of larger SPs is more hindered by the pore structure, thereby enhancing the deposition of adsorbed HMs. A similar trend was observed when the injection concentration of Pb^{2+} was increased to $C_P = 200$ ng/ml (i.e., in Stage 2, $V_P > 2.75$). In this case, the effluent concentrations of Pb^{2+} corresponding to the cases without SPs and with SPs of $D_{50} = 5.1$, 17.1, and 24.7 μm were 127.9, 119.2, 81.2, and 71.2 ng/ml, respectively. This occurs because the porous matrix alone has a limited attachment-capacity for Pb^{2+}, leading to higher effluent concentrations. In contrast, when SPs are present, Pb^{2+} adsorbed onto their surfaces can be retained within the media via SPs deposition, resulting in lower effluent levels. Figure 4.12a and b reveals that higher SPs injection concentrations strengthen the inhibitory effect on Pb^{2+}, corresponding to a further reduction in their effluent concentration. At a relatively high seepage velocity (e.g., v = 0.260 cm/s; not shown due to space constraints), the transport behavior of Pb^{2+} was analogous to that at

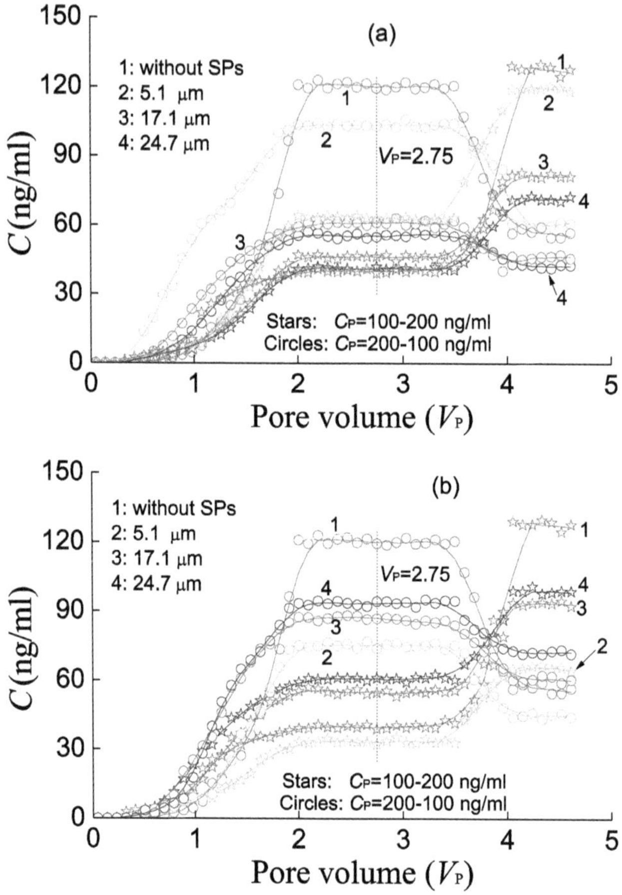

FIGURE 4.12 BTCs of Pb^{2+} under different Pb^{2+} concentrations. (a) $C_S = 0.5$ mg/ml. (b) $C_S = 2$ mg/ml.

$v = 0.087$ cm/s, although with an overall tendency for effluent concentrations to increase with greater seepage velocity.

4.4 COTRANSPORT PROCESS AT DIFFERENT TEMPERATURES

4.4.1 Experimental Description

Experiments were conducted to investigate the solitary transport of Pb^{2+} or Cd^{2+} (Bai et al., 2021c), as well as their cotransport with SPs (i.e., a kind of silica powders; commercial, natural, irregular SiO_2 particles). The concentration levels for Pb^{2+} solutions were established at $C_M = 100, 300,$ and 500 μg/ml, while for Cd^{2+} solutions, the levels were set at $C_M = 100, 300,$ and $500\,\text{ng/ml}$. Considering the potential temperature range of discarded solutions in practical engineering applications, varying temperatures (e.g., T = 20 °C, 40 °C, and 60 °C) were also incorporated into the experimental design. The injection concentration of SPs remained constant at $C_{\text{inj}} = 2$ mg/ml.

The experimental procedure commenced 30 min after both the water bath and the temperature-controlled water tank had stabilized at the target temperature to maintain a constant temperature, thereby ensuring a consistent thermal environment during the entire test. As a single pulse-injection pattern, a solution volume of $V_{inj} = 30$ ml (Pb^{2+}, Cd^{2+}, or SPs alone or a mixture of Pb^{2+} or Cd^{2+} with SPs) was injected into the specimen column over a short period of 2 s.

4.4.2 Transport Process of HMs

Figure 4.13 presents the BTCs for Pb^{2+} under varying conditions of temperature, injection concentrations, and seepage velocities. Comparable trends are evident in the BTCs obtained for Cd^{2+}. A significant disparity is observed between the peak relative concentration (C_R) of Pb^{2+} and Cd^{2+} in the effluent, which can be attributed to the large difference in concentration between the Pb^{2+} and Cd^{2+} injection solutions (Pb^{2+}: $C_M = 100, 300, \text{and } 500\ \mu g/ml$; Cd^{2+}: $C_M = 100, 300, \text{and } 500\, ng/ml$) and in their adsorption and desorption in the porous medium. Nevertheless, the experimental results show that at the same C_M, temperature appears to not affect the transport of HMs in the porous medium. For instance, the peak C_R values of Pb^{2+} for T = 20 °C, 40 °C, and 60 °C are very close ($C_R = 0.058 - 0.060$) at a seepage velocity of v = 0.087 cm/s and an injection concentration of $C_M = 100\ \mu g/ml$, whereas for Cd^{2+}, the same phenomenon can also be observed ($C_R = 0.226 - 0.238$) at v = 0.087 cm/s and $C_M = 100$ ng/ml. However, Pb^{2+} is more likely to be attached than Cd^{2+} due to its higher adsorption on the negatively charged matrix of

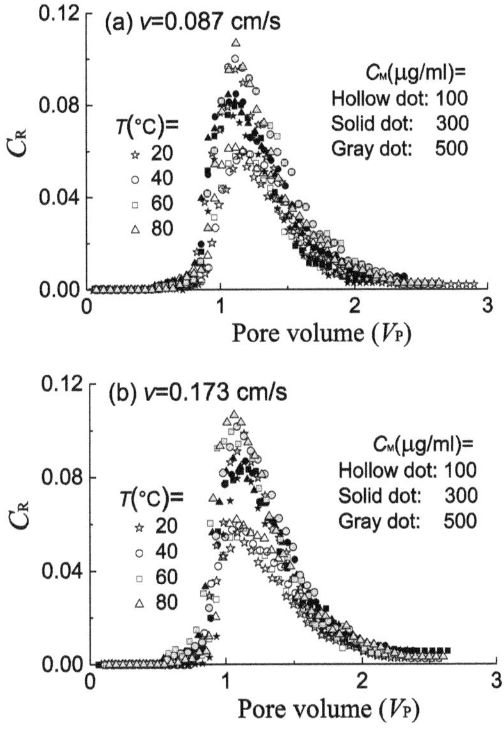

FIGURE 4.13 BTCs of Pb^{2+} at different temperatures. (a) v = 0.087 cm/s. (b) v = 0.173 cm/s.

the porous medium (the main mineral constituent is SiO_2). Consequently, this enhanced adsorption leads to a lower peak C_R for Pb^{2+} in the leachate, as illustrated in Figure 4.13.

Figure 4.13 displays a pronounced tailing phenomenon in the later stages of the BTCs, which signifies the desorption of previously adsorbed HMs. A key observation is that the BTCs recorded at seepage velocities of v = 0.087, 0.173, and 0.260 cm/s are virtually identical, suggesting that seepage velocity has a minimal impact on the transport of individual HMs. In an acidic environment, a fraction of the HMs remains dissolved and mobile within the aqueous phase, while the rest is adsorbed onto the soil matrix. A portion of these adsorbed HMs can subsequently be released back into the solution when the aqueous HMs concentration becomes sufficiently low. The experimental data clearly indicate that, within the range of 20–80°C, temperature variations do not produce significant effects on HMs transport, thereby yielding nearly overlapping BTCs across different temperatures. In contrast, as the injection concentration of HMs increases, the peak C_R increases accordingly. For instance, Figure 4.13 indicates that an increase in the injection concentration of Pb^{2+} at v = 0.087 cm/s causes an increase in the average peak C_R and actual concentration ($C = 0.279 \rightarrow 1.146 \rightarrow 2.352$ μg/ml) of Pb^{2+} at the four temperatures. Test results also reveal that an increase in the injection concentration of Cd^{2+} from $C_M = 100$ ng/ml to $C_M = 500$ ng/ml at v = 0.087 cm/s leads to an increase in the average peak C_R from $C_R = 0.233$ to $C_R = 0.478$.

4.4.3 Cotransport Processes of HMs and SPs

Figure 4.14 indicates that seepage velocity exerts no substantial influence on the transport of Pb^{2+} when SPs are present, as evidenced by the nearly overlapping BTCs. These curves also exhibit similar tailing phenomena, which correspond to the desorption of deposited HMs. The peak C_R of Pb^{2+} in the presence of SPs is higher than that in the absence of SPs (Figure 4.13). For instance, in the presence of SPs ($D_{50} = 2.8$ μm, $C_M = 100$ μg/ml, and T = 20 °C), the peak C_R of Pb^{2+} is 14.8 times that in the absence of SPs, indicating that SPs can significantly facilitate the transport of Pb^{2+}. Similarly, the peak C_R of Cd^{2+} in the presence of SPs ($D_{50} = 2.8$ μm, $C_M = 100$ ng/ml, and T = 20 °C) is 2.9 times that in the absence of SPs, which is lower than that of Pb^{2+}. Thus, the transport of Cd^{2+} is less facilitated by SPs than that of Pb^{2+}, which can be attributed to the weak adsorption of Cd^{2+} on SPs. In general, the higher the amount of HMs adsorbed on SPs, the more significant the SP-facilitated transport of HMs. However, it is important to note that the presence of SPs with a large particle size may inhibit the transport of Cd^{2+} at high temperatures and injection concentrations (e.g., $D_{50} = 18.7$ μm, T = 60 °C, $C_M = 500$ ng/ml); thus, the concentration of Cd^{2+} in the leachate in the presence of SPs ($C_R = 0.398$) is slightly lower than that in the absence of SPs ($C_R = 0.463$).

Figure 4.14 demonstrates a significant decrease in the peak relative C_R of HMs with rising temperature, a trend that aligns with the BTCs observed for SPs in Figure 4.15. This correspondence indicates that the transport of SPs exerts a considerable interdependent influence on HMs transport, primarily mediated through the adsorption of HMs onto the SPs. Consequently, the concentration of HMs (especially for Pb^{2+}) in the effluent is

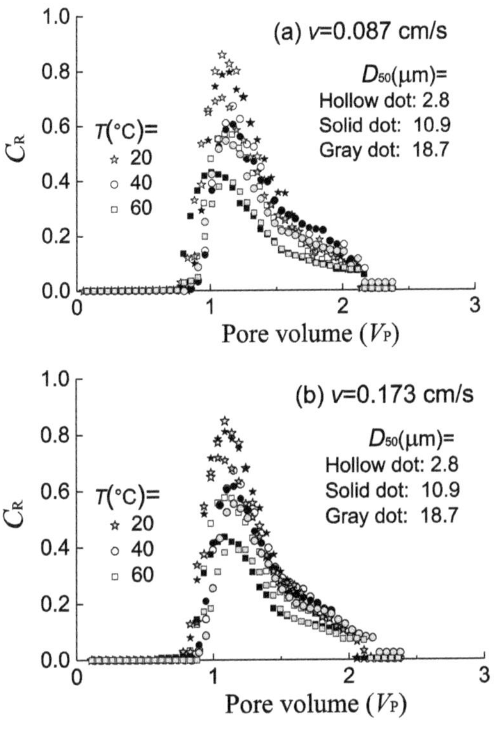

FIGURE 4.14 BTCs of Pb^{2+} facilitated by SPs ($C_M = 100$ mg/ml). (a) $v = 0.087$ cm/s. (b) $v = 0.173$ cm/s.

critically influenced by the concurrent concentration of SPs. Furthermore, due to the intricate physical and chemical interactions occurring between the porous medium matrix, the HMs and the SPs can substantially retard the particles themselves. This retardation, amplified at elevated temperatures, leads to an increased attachment coefficient and ultimately manifests as a lower peak C_R in the BTC.

A comparison of Figures 4.13 and 4.14 indicates that the introduction of SPs elevates the peak C_R of HMs and induces an earlier arrival of Pb^{2+}, as evidenced by a smaller V_P corresponding to the peak C_R. This observation confirms that SPs can substantially enhance the transport of Pb^{2+}. The underlying mechanism involves the adsorption of positively charged HMs (e.g., Pb^{2+}) onto the surfaces of negatively charged SPs, which possess a significant attachment-capacity. Acting as a mobile solid phase alongside the immobile solid medium and the aqueous phase, these particles-bound HMs are carried with the flow. Consequently, Pb^{2+} adsorbed onto SPs can migrate through the subsurface at a velocity comparable to or exceeding that of the dissolved phase, a phenomenon further amplified by the size exclusion effect.

Figure 4.15 displays the BTCs of SPs, both with and without the presence of Pb^{2+} at $C_M = 100$ μg/ml and $C_{inj} = 2$ mg/ml. A clear reduction in the peak values of SPs (C_R) is observed when Pb^{2+} is present, particularly for SPs with a larger particle size (e.g., $D_{50} = 18.7$ μm). This demonstrates a reciprocal coupling effect where Pb^{2+} also influences the transport dynamics of the SPs themselves.

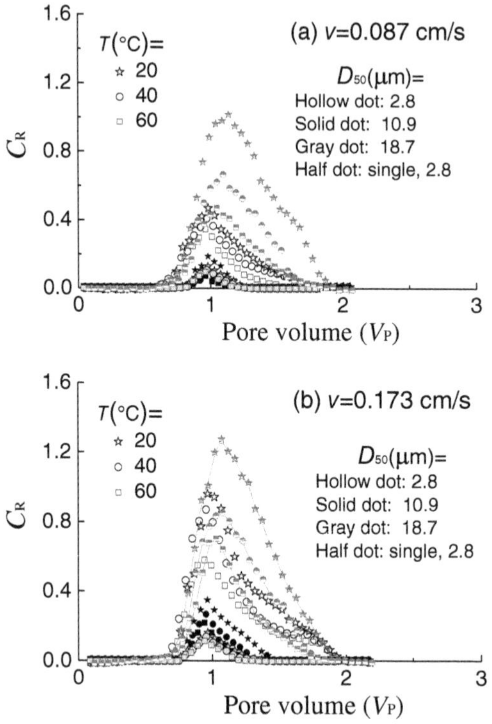

FIGURE 4.15 BTCs of SPs during cotransport with Pb^{2+}. (a) v = 0.087 cm/s. (b) v = 0.173 cm/s.

Both the peak values of HMs and that of SPs exhibit a decline as the particle size of SPs increases (Figures 4.14 and 4.15). This trend occurs because larger SPs are more susceptible to physical retention at pore throats within the soil matrix. This entrapment elevates the attachment coefficient for the SPs and can delay the arrival of their peak concentration. Consequently, HMs adsorbed onto these retained SPs are also deposited, underscoring a strong interdependent transport mechanism between the particles and the HMs. This coupling explains the observed inverse relationship for Pb^{2+}: within the tested concentration range ($C_M = 100 - 500$ μg/ml), its peak C_R decreases as the injection concentration rises, which contrasts with its transport behavior in the absence of SPs. The rationale is that a higher Pb^{2+} concentration leads to greater adsorption onto the SPs, which in turn enhances the electrostatic attraction between the now more positively charged SPs and the negatively charged porous medium matrix. This promotes the deposition of the SPs themselves onto the solid matrix, reducing the number of mobile particles and thereby lowering the concentration of particle-facilitated Pb^{2+} in the effluent. In contrast, the peak C_R of Cd^{2+}, which adsorbs only weakly onto SPs, increases with its injection concentration. Here, the transport of Cd^{2+} by itself is dominant. For instance, increasing the injection concentration of Cd^{2+} from $C_M = 100$ ng/ml to $C_M = 500$ ng/ml at v = 0.087 cm/s, $D_{50} = 2.8$ μm, and T = 20 °C results in an increase in the peak C_R from $C_R = 0.667$ to $C_R = 1.070$. A further comparison of Figures 4.14 and 4.15 indicates that the peak C_R for

SPs occurs earlier than that for the HMs. This earlier arrival highlights the accelerated transport of the SPs through the porous medium, a result of the size exclusion effect.

4.5 DEPOSITION CHARACTERISTICS ALONG THE TRANSPORT DISTANCE

4.5.1 Experimental Description

4.5.1.1 Experimental Apparatus and Materials

The experimental setup utilized a Plexiglas apparatus, as illustrated in Figure 4.16, designed to provide an extended seepage path (Bai et al., 2020c). The inner diameter of the hollow cylindrical chamber is $D = 8$ cm, and the length is $L = 90$ cm. The length-to-diameter ratio is $L/D = 11.25$, which can be regarded as a one-dimensional sand column test. To simplify the process of packing the sand specimen and to enable posttest sampling for measuring deposition concentrations, the cylindrical chamber was constructed in three modular sections. Each section is 30 cm in length. These segments are securely joined using bolts and sealed with rubber rings to ensure the integrity of the assembly and prevent leakage.

The test solution (Pb^{2+}, SPs, or Pb^{2+} mixed with SPs) in the water tank was pumped into the column at a constant velocity using a peristaltic pump, and the measurement precision is 0.07 ml/min. A digital flowmeter installed at the column inlet was used to monitor the seepage velocity. To ensure homogeneity, the mixed solutions were kept in suspension within the water tank by a motorized stirrer. The flow was directed vertically from the top to the bottom of the column. In this configuration, gravitational force provided additional

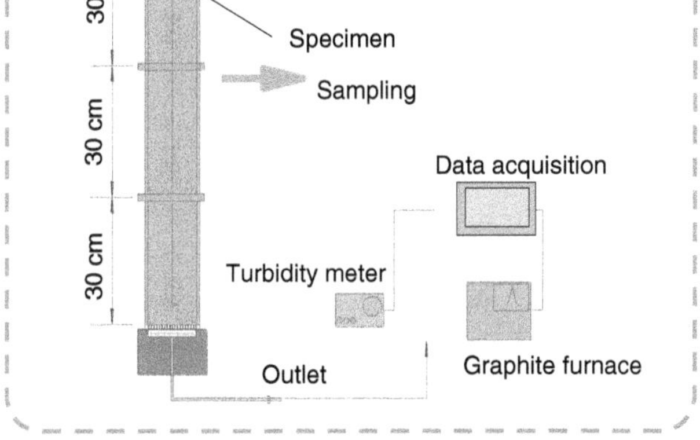

FIGURE 4.16 Experimental equipment.

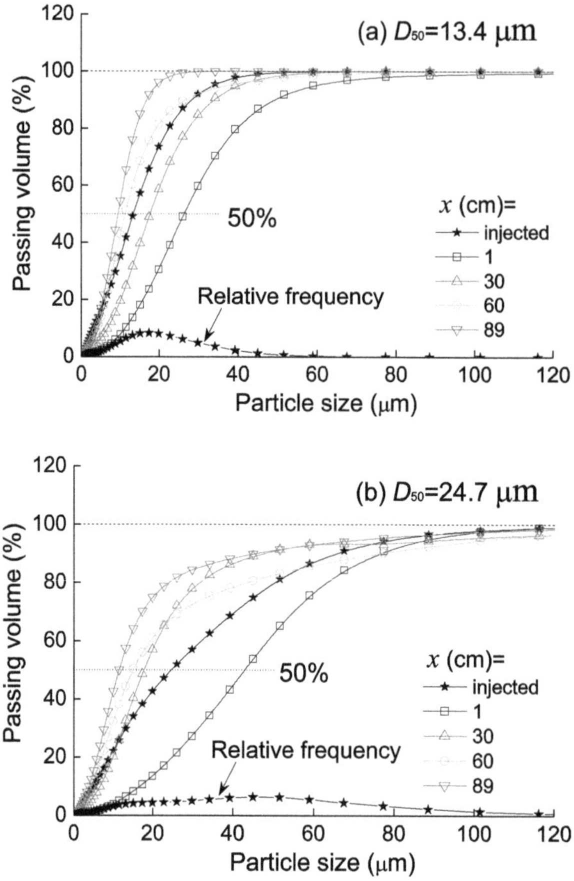

FIGURE 4.17 PSD of the injected and deposited SPs ($v = 0.260$ cm/s, $C_M = 100$ ng/ml, $C_{inj} = 4$ mg/ml; wet method). (a) $D_{50} = 13.4$ μm. (b) $D_{50} = 24.7$ μm.

acceleration for the transport of SPs, as it acted in the same direction as the seepage force exerted by the water flow on the surface of SPs.

The porous medium employed consisted of natural quartz sand ($\rho_s = 2.65$ g/cm^3), prepared by mechanical crushing. The sand particles were uniformly sized within the range of 1–2 mm, with a mean grain d_g of 1.5 mm. Since Pb^{2+} in solution typically induces a slightly acidic condition, the Pb^{2+} solution was adjusted to pH of 5.5 using small volumes of 0.01 mol/l NaOH or HNO_3 solution. The SPs were two types of commercial spherical particles produced by the dry grinding of high-quality, naturally colored quartzite, which is insoluble in both acidic and alkaline solutions. Figure 4.17 shows the measured PSDs by laser diffraction (wet method). The two particle size ranges were d =1–60 μm and 1–120 μm with median diameters of D_{50} = 13.4 and 24.7 μm, respectively. The injected SPs actually had a very wide PSD, which reflects the situation in practical engineering, whereby the mobilized solid particles may be very inhomogeneous. The two SPs were red and black in color, and both had a density of $\rho_s = 2.2$ g/cm^3.

4.5.1.2 *Column Experiments*

The quartz sand was incrementally packed into the column using a layered method. Based on the total quality required, the sand was divided into ten parts, each forming a layer approximately 9 cm in height after placement. The sand was carefully stirred with a glass rod and then compacted to a specified thickness using a tamping hammer. This layered procedure was repeated, with deionized water added and maintained at a level 2 cm above the sand surface throughout the packing process to ensure full saturation. The resulting saturated sand column had a final measured porosity of n = 41.7%.

The injection concentrations of SPs were established at $C_{\text{inj}} = 0.5$, 2, and 4 mg/ml. Before each experiment, the suspension in the water tank was homogenized for approximately 30 min using a motorized stirrer to ensure adequate contact time and uniform distribution. The Pb^{2+} solution concentrations were selected as $C_{\text{M}} = 100$, 200, and 400 ng/ml. Seepage velocities were set at v = 0.087 and 0.260 cm/s, corresponding to Reynolds numbers $\text{Re} = 1.45 - 4.34 < 10$, which satisfied the laminar flow condition. This range represents a relatively high seepage velocity, making the associated shear stresses acting on the SPs a significant factor for consideration. Such velocities apply to studies of suspended substance transport in coarse-grained strata and to active remediation technologies for heavy metal pollutants in porous media. All tests were conducted under a controlled temperature of T = 22±0.2 °C.

In this experimental procedure, 15 ml of a 0.05 mol/l EDTA disodium salt solution was employed as the extraction agent, which demonstrated high extraction efficacy (>90%). The leachate was subsequently agitated for 4 h using a constant-temperature oscillator. Following this, Pb^{2+} and SPs were separated via a desktop low-speed centrifuge. Upon concluding the transport tests, representative samples were extracted from various segments of the column, with each sampled layer having a thickness of approximately 1 cm. The concentrations of deposited Pb^{2+} and SPs within these samples were measured. The deposition patterns of the SPs were examined with a 5-megapixel metallographic microscope, which readily distinguishes the colored SPs from the white quartz sand matrix.

4.5.2 Transport Process of HMs

Figure 4.18 presents the BTCs for Pb^{2+} alone under varying injection concentrations and seepage velocities, in a system where SPs were absent. A clear trend is observed where the plateau concentration increases with higher injection C_{M}, while the time required to reach this steady state is correspondingly reduced. Obviously, the detection time of Pb^{2+} at the outlet of the column remains the same for a given seepage velocity (e.g., approximately 1 V_{P} for v = 0.260 cm/s and approximately 0.8 V_{P} for v = 0.087 cm/s), while it is advanced as the seepage velocity increases. Additionally, the effluent concentration shows a slight elevation with increasing seepage velocity ($v = 0.087$ cm/s $\rightarrow$ 0.260 cm/s). These collective observations indicate a comparatively minor attachment effect under high flow conditions.

For test results employing the pulse-injection method over a period of $t_{\text{inj}} = 2$ s at elevated concentration (Pb^{2+}, $C_{\text{M}} = 300 - 500\,\mu\text{g/ml}$), the BTCs at v = 0.087 and 0.260 cm/s are nearly identical. This suggests that seepage velocity has a minimal influence on the pore volume V_{P} corresponding to the peak Pb^{2+} concentration within the tested seepage

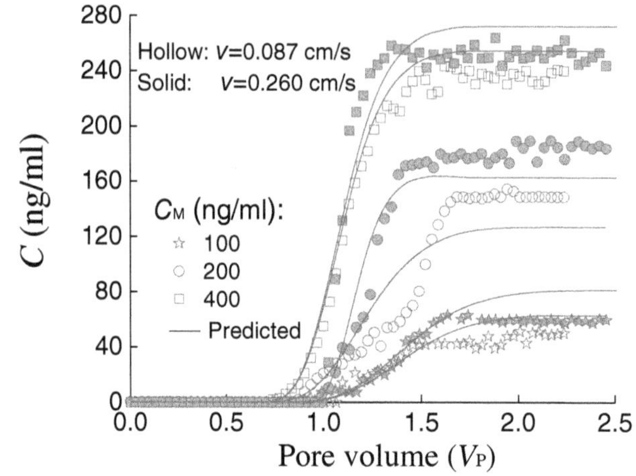

FIGURE 4.18 BTCs of Pb^{2+} in the absence of SPs.

range. Under this injection mode, soluble Pb^{2+} move almost simultaneously with the water flow, and the transport distance is directly proportional to the seepage velocity. This observed behavior is slightly different from the test results obtained by the so-called continuous injection pattern (i.e., long-term injection; Figure 4.18). The injection periods were approximately 960 and 360 s for v = 0.087 and 0.260 cm/s, respectively (herein, $C_M = 100 - 400\,\text{ng/ml}$). In the transport of HMs, a fraction remains dissolved and mobile, while another portion is adsorbed onto the solid matrix. Compared to the pulse-injection case, the continuous injection pattern exhibits a more pronounced deposition and subsequent release process, particularly as the injection concentration decreases.

In contrast to the effluent concentration of Pb^{2+} observed in the absence of SPs (Figure 4.18), the steady-state effluent concentration in the presence of SPs (Figures 4.19 and 4.20) exhibits greater complexity. Specifically, the introduction of SPs can either enhance Pb^{2+} transport—typically with smaller SPs—or inhibit it, as often seen with larger particles. The net effect appears to be closely dependent on several interacting factors: the injected Pb^{2+}, seepage velocity, and concentration and size of injected SPs. For instance, the steady concentration of Pb^{2+} in the presence of SPs ($C_{inj} = 0.5$ mg/ml ; Figure 4.19) for a slightly smaller particle size (e.g., $D_{50} = 13.4\ \mu\text{m}$) and at a lower injection concentration of Pb^{2+} (e.g., $C_M = 100$ ng/ml) and higher seepage velocity (e.g., v = 0.260 cm/s) is nearly 1.1 times the steady concentration in the absence of SPs (see Figure 4.18).

As the injection concentration of Pb^{2+} increases (e.g., $C_M = 400$ ng/ml), the presence of SPs begins to inhibit Pb^{2+} transport, primarily due to the accelerated deposition of SPs. For instance, under conditions of a moderately small particle size (e.g., $D_{50} = 13.4\ \mu\text{m}$) at $C_{inj} = 0.5$ mg/ml (Figure 4.19), the steady-state concentration of Pb^{2+} with SPs present is only 0.5 times that measured without SPs (Figure 4.18). Due to the addition of $Pb(NO_3)_2$, Pb^{2+} adsorption onto SPs reduces the repulsive force between SPs and the solid matrix due

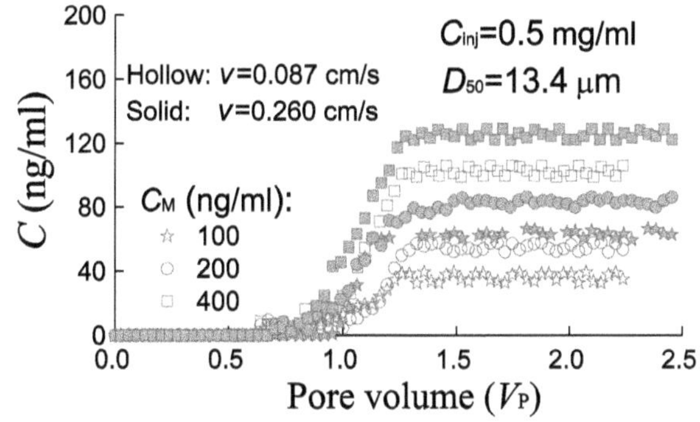

FIGURE 4.19 BTCs of Pb^{2+} in the presence of SPs for $D_{50} = 13.4\ \mu m$.

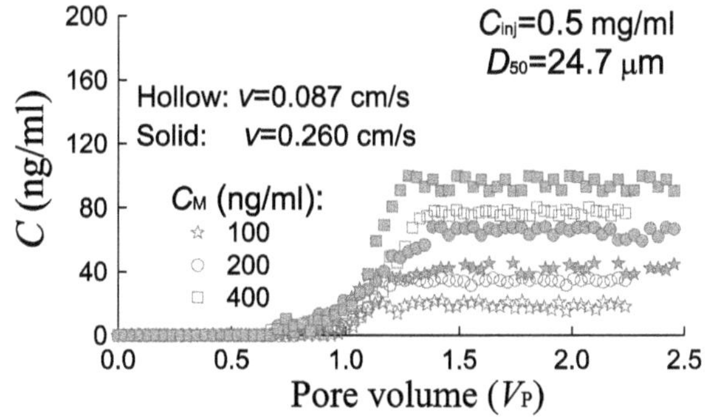

FIGURE 4.20 BTCs of Pb^{2+} in the presence of SPs for $D_{50} = 24.7\ \mu m$.

to the decrease in absolute zeta potential of the surface charge, resulting in an increase in the deposition amount of SPs onto the solid matrix. Consequently, the coupling effect between positively charged HMs, like Pb^{2+}, and the SPs must be considered, a phenomenon linked to the compression of the electrical double layer on the SPs surfaces (i.e., a decrease in surface potential energy). Overall, the facilitative effect of SPs on the transport of Pb^{2+} is amplified with higher injection concentrations but is markedly reduced with larger particle sizes, as is evident from the differences between Figure 4.19 and Figure 4.20.

4.5.3 Distribution of HMs and SPs along the Transport Distance

Figure 4.21 depicts the measured deposition concentration (namely, attachment concentration, σ_M) of Pb^{2+} along the transport path following the completion of transport tests, specifically for co-injected SPs with a median particle size of $D_{50} = 13.4\ \mu m$, including the results in the absence of SPs. The reported deposition amount comprises Pb^{2+} adsorbed directly onto the solid matrix as well as Pb^{2+} adsorbed onto SPs that were subsequently deposited. The data indicate that the Pb^{2+} deposition concentration on the solid matrix declines sharply with increasing transport distance. When the transport distance is reduced to $x = 90$ cm (i.e., the length of the sand column), the deposition concentrations are already

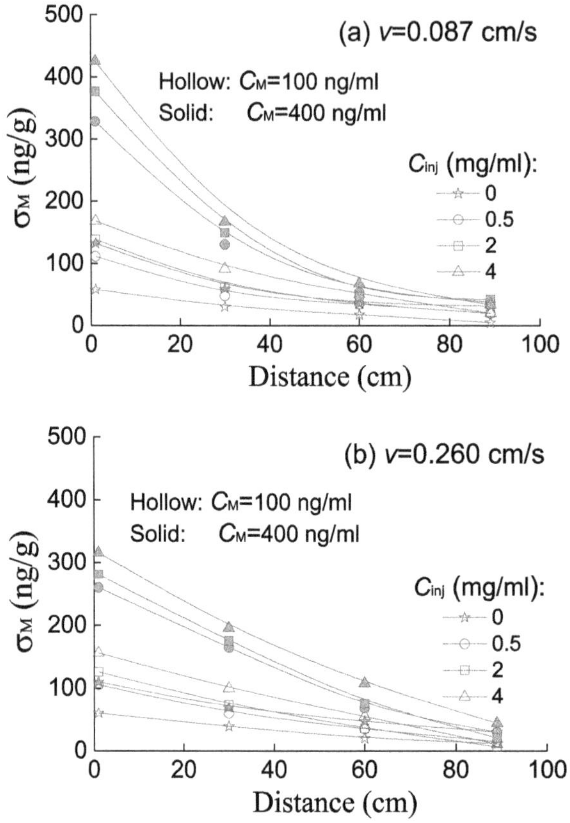

FIGURE 4.21 Deposition concentrations of Pb^{2+} along the transport distance. (a) $v = 0.087$ cm/s. (b) $v = 0.260$ cm/s.

very low. In addition, with increasing seepage velocity (e.g., $v = 0.087–0.260$ cm/s), the Pb^{2+} deposition concentration also notably decreases with decreasing the concentration of SPs in the water flow (see Figure 4.21). In the absence of SPs, Pb^{2+} deposition is significantly lower, particularly at low Pb^{2+} injection concentrations. Consequently, variations in seepage velocities lead to only minor differences in deposition amounts under these conditions, and the influence of electrostatic adsorption between HMs and the solid matrix on deposition becomes a nonnegligible factor.

Comparable outcomes were observed for SPs with a larger median diameter ($D_{50} = 24.7\ \mu m$); these data are not presented here due to spatial constraints. Overall, as the particle size of SPs increases, the deposition amount of the SPs themselves rises substantially, whereas the deposition amount of Pb^{2+} shows an opposite, decreasing trend. This indicates that beyond a certain particle size threshold, the attachment quantity of the SPs for HMs is diminished. Consequently, the facilitative effect of SPs on the transport of Pb^{2+}, or their inhibitory effect when deposition is enhanced, becomes significantly less pronounced.

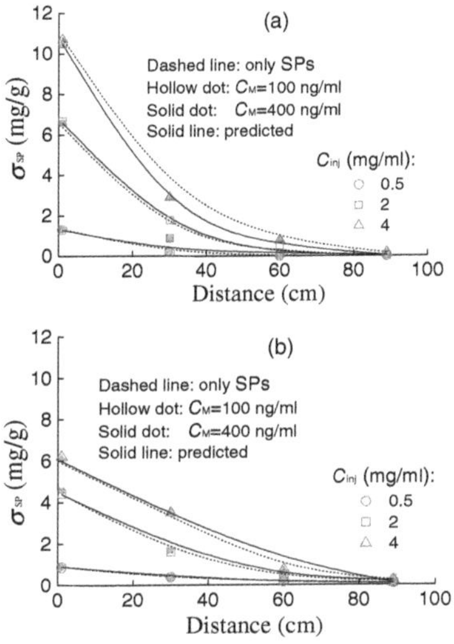

FIGURE 4.22 Deposition concentrations of SPs along the transport distance. (a) $v = 0.087$ cm/s. (b) $v = 0.260$ cm/s.

Figure 4.21 indicates that Pb^{2+} deposition rises with high injection concentrations. Under these conditions, a greater amount of Pb^{2+} is adsorbed onto the surface of SPs and subsequently migrates or is deposited as coupled Pb^{2+} and SPs with the flowing water, which is reflected by the deposition concentration (σ_{SP}) of SPs with the transport distance, as shown in Figure 4.22 $\left(D_{50} = 13.4\ \mu\text{m}\right)$. It should be noted that the concentration of the injected Pb^{2+} in this test is arguably low, and thus, most Pb^{2+} will be adsorbed onto the surface of SPs, forming a combination of coupled Pb^{2+} and SPs. However, when the Pb^{2+} injection concentration is substantially increased, Pb^{2+} no longer exists solely in this combined form. A significant portion remains as free ions in the aqueous phase and is transported directly by the water flow. Under these high concentration conditions, as demonstrated in Figures 4.19 and 4.20, the Pb^{2+} concentration measured in the leachate shows a marked increase.

The adsorption of HMs onto SPs alters their dielectric properties, rendering their surfaces more positively charged. This modification enhances the adsorption of SPs onto the porous medium matrix and increases their probability of deposition. The Pb^{2+} injection concentration appears to have minimal discernible impact on SPs deposition during transport (Figure 4.22). This is likely attributable to the relatively low Pb^{2+} concentration employed. At such levels, the adsorption of a limited quantity of HMs is insufficient to substantially modify the surface charge characteristics of SPs. Consequently, the transport behavior of SPs themselves can still be adequately described by the transport theory of a single, homogeneous suspension. In contrast, the coupling effect of SPs on HMs transport must be considered. Hence, HMs will migrate and be deposited in two forms, that is, dissolved in the flowing water and absorbed onto the SPs.

The spatial distributions of accumulated SPs along the transport distance have been determined, as represented by the solid curves in Figure 4.22. Figure 4.22 also includes the deposition profiles from experiments involving only SPs transport. Clearly, the reaction rate constant (i.e., λ of Eq. (3.8) in Chapter 3) of SPs interacting with the solid matrix is far larger than that of Pb^{2+} interacting with the solid matrix, which essentially reflects a more notable hysteretic effect of the SPs attached to the matrix by the straining mechanism. It can be inferred that the transport of metallic cations is intimately linked to the movement of SPs, owing to the strong adsorption affinity of these positively charged HMs for SPs surfaces. In other words, the magnitude of the mutual influence between Pb^{2+} and SPs on their transport and deposition characteristics depends on the physic-chemical properties.

4.5.4 Deposition Rate and Particle Size Distributions

The ratio of the mass of a migrating substance that has passed through a given cross section x at time t to the total mass injected into the porous medium is defined as follows:

$$M_b = \frac{\int_0^t C(x,t) \cdot u \cdot (nA) \cdot dt}{m} \tag{4.1}$$

where M_b (%) is the penetration rate (i.e., the recovery rate), $C(x,t)$ is the suspended substance concentration of the coordinate x at time t, A is the cross-sectional area of the sand column, and m is the total mass of the transport substance injected into the porous medium.

The deposition rate (M_a) is defined as $M_a = 1 - M_b$ (%), which denotes the amount of suspended substances deposited in the range of (0, x) in the column.

Based on the experimental results presented in Figure 4.21, the deposition rate of Pb^{2+} throughout the entire column relative to the injection concentration of SPs, C_{inj}, can also be expressed by Eq. (4.1) (see Figure 4.23). The data clearly show that as the injection concentration of SPs is raised (e.g., $C_{inj} = 0$, 0.5, 2, and 4 mg/ml), the amount of deposited Pb^{2+} gradually increases. Here, $C_{inj} = 0$ indicates the Pb^{2+} deposition rate in the absence of SPs. Conversely, when the Pb^{2+} injection concentration is elevated ($C_M = 100$, 400 ng/ml), the quality of deposited Pb^{2+} diminishes. This occurs because a portion of the HMs dissolves into the aqueous phase and is transported with the flowing water. Consequently, the deposition rate of HMs shows a slight decrease under higher seepage velocities, and the dissolved fraction progressively becomes the dominant transport pathway as the injection concentration increases.

When the particle size of SPs increases (e.g., $D_{50} = 13.4 \rightarrow 24.7\ \mu m$), the overall deposition mass of the particles rises. However, this larger particle size corresponds to a reduction in specific surface area, which diminishes the attachment-capacity of the SPs for Pb^{2+}. Consequently, the deposition rate of Pb^{2+} declines. Furthermore, the positive correlation between Pb^{2+} concentration and injection concentration of SPs becomes less pronounced, leading to a downward trend in the Pb^{2+} deposition rate (Figure 4.23a).

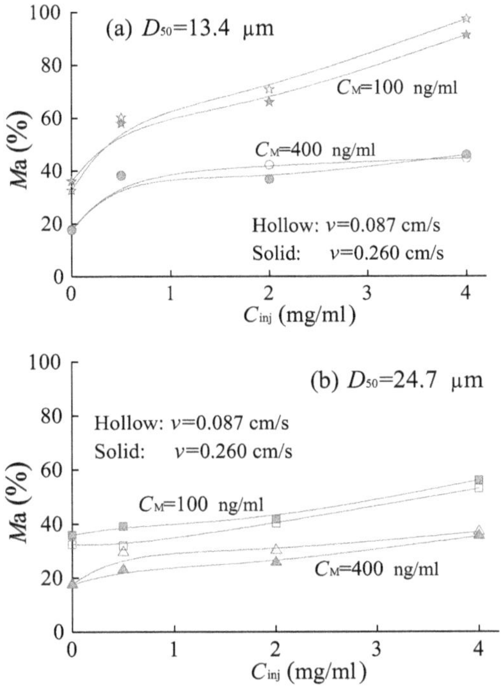

FIGURE 4.23 Deposition rate of Pb^{2+} with the injection concentration of SPs. (a) $D_{50} = 13.4$ μm. (b) $D_{50} = 24.7$ μm.

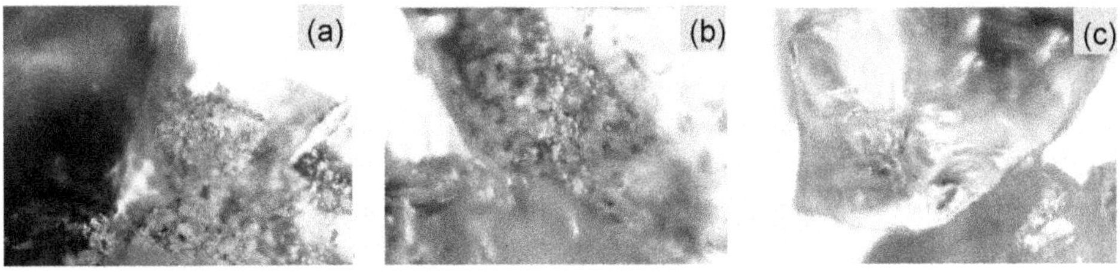

FIGURE 4.24 Microstructure photos of the deposited SPs for v = 0.087 cm/s. (a) x = 1 cm. (b) x = 30 cm. (c) x = 60 cm.

For SPs with a smaller median diameter (e.g., $D_{50} < 10$ μm), the preferential flow induced by the size exclusion effect can accelerate the transport of HMs (e.g., Pb^{2+}) along with the SPs. For SPs with larger median diameter (e.g., $D_{50} = 13.4$ and 24.7 μm), the transport of these larger particles is subject to significant deposition due to the pore blocking effect of the porous medium, and, eventually, HMs transport will be inhibited rather than promoted. The microstructure photos after the transport tests (Figure 4.24; $D_{50} = 13.4$ μm and C = 2 mg/ml) reveal that the SPs with larger particle sizes are first deposited within a relatively short distance from the injection surface of the sand column (e.g., x = 1 cm), while the deposition amount of SPs gradually decreases with increasing distance. The amount of deposited particles then gradually diminishes with increasing distance, followed by a sharp decline at specific intervals (e.g., x = 30 and 60 cm). This deposition pattern aligns with the concentration profiles along the sand column (Figure 4.22).

The so-called straining effect is sensitive to the ratio of the solid particle diameter D_{50} to the porous medium grain diameter d_g. In the experiments, $D_{50}/d_g = 0.0089 - 0.0165$ (i.e., $D_{50} = 13.4$ and 24.7 μm and $d_g = 1.5$ mm), and this ratio is greater than the generally considered critical ratio (e.g., 0.002–0.008). Consequently, particle retention via straining becomes the dominant mechanism, overshadowing any preferential flow effects. Large-sized SPs can lead to pore narrowing (see Figure 4.24), which enhances the deposition probability of SPs by the straining effect, especially at a low velocity (e.g., v = 0.087 cm/s; Figure 4.24). With increasing seepage velocity (e.g., v = 0.087–0.260 cm/s), the deposition amount of SPs at the same distance decreases, and they advance to locations farther away from the injection end. That is, at a low seepage velocity, the porous medium captures most of the injected larger-sized SPs at the entrance. At a high seepage velocity, the water flow carries the particles deeper into the porous medium, producing more gradual changes in the deposition profile. Naturally, at any given seepage velocity, the transport distance of larger SPs (e.g., $D_{50} = 24.7$ μm) remains constrained to a relatively narrow range within the porous medium.

Figure 4.17 presents the PSD of the two selected SPs (i.e., $D_{50} = 13.4$ and 24.7 μm) extracted from different sections of the sand column after completing the transport tests. Figure 4.17 shows that with increasing distance from the injection end, the particle size of the deposited SPs gradually decreases, exhibiting a clear particle-separation characteristic due to the flowing water. For the injected SPs of $D_{50} = 13.4$ μm with a wide range of d =1–60 μm, the median diameters are $D_{50} = 26.4$, 17.6, 11.3, and 9.5 μm when x = 1, 30, 60, and 89 cm, respectively; while for the injected SPs of $D_{50} = 24.7$ μm with a range of d = 1–120 μm, the median diameters are $D_{50} = 42.7$, 17.8, 15.0, and 11.5 μm, respectively. Evidently, near the injection end (e.g., x = 1 cm), the median diameter of the deposited SPs can exceed that of the injected population, then gradually decreases along the transport distance, and finally, the median diameter will be smaller than that of the injected SPs near the outflow (e.g., x = 89 cm).

CHAPTER 5

Transport of Heavy Metal Ions in Unsaturated Soils

5.1 MAIN CONTENT OVERVIEW

Static attachment–detachment (or adsorption–desorption) experiments of heavy metal ions (HMs) are conducted using silty soil as a porous medium. The adsorption mechanism of representative HMs is analyzed based on the physical properties and microstructure features, including functional groups, chemical elements, and mineral compositions. The influence of temperature, types of HMs, and the concentration on the attachment–detachment characteristics are explored. By means of a one-dimensional unsaturated soil column permeability test, the water saturation process and the transport mechanism of HMs are examined during the injection of HMs-contaminated water. The impact of the concentration and HMs types on both soil moisture content and the spatial distribution of HMs are evaluated systematically. Furthermore, the physical nature of the attachment–detachment model is elucidated, taking into account the adsorption hysteresis effect.

As a particular aspect of this research, a pore-scale simulation approach is employed to predict the thermal conductivities of unsaturated porous media via the smoothed particle hydrodynamics (SPH) method. The correlation between the normalized thermal conductivity and saturation degree is subsequently analyzed using an empirical model. This section outlines the reconstruction process of unsaturated granular soils at the pore-scale along with the simulation procedure, and validates the accuracy of the simulation results through numerical experiments on ideal media. Furthermore, it details simulation experiments conducted in 3D granular porous media and examines how the thermal conductivity varies with saturation degree.

5.2 TRANSPORT PROCESS OF HMs

5.2.1 Typical Tests

5.2.1.1 Test Soil

The experimental silty soil was sourced from Lingshou County, Hebei Province, China. The limit water contents, such as the liquid limit, plastic limit, and plasticity index of the silty

DOI: 10.1201/9781003783749-5

soil, were 20.9%, 12.2%, and 8.7, respectively. The particle specific gravity measured 2.71, while the maximum dry density and optimum water content were found to be 1.51 g/cm^3 and 16%, respectively. Particle size distribution (PSD) was assessed with a laser scattering particle size analyzer, yielding a range of 0.1–100 μm and an average particle size of 30.5 μm. The microstructure images of raw silty soil (SEM) are shown in Figure 5.1a, and its specific surface area is 28.32 m^2/g. The mineral compositions of the silty soil are quartz (24%), illite (18%), kaolinite (6%), plagioclase (15%), microcline (14%), calcite (2%), hematite (2%), and montmorillonite (19%). Through the EDS energy spectrum test, the main constituent elements of the silty soil were obtained: O (48.1%), C (39.7%), Si (5.0%), Al (3.8%), Fe (1.4%), Ca (0.2%), K (0.2%), and Cl (0.1%).

5.2.1.2 Heavy Metal Ions

The Pb^{2+}, Cu^{2+}, and Cd^{2+} were chosen as representative HMs pollutants. The Pb^{2+} solution, prepared from $Pb(NO_3)_2$, was a colorless, transparent liquid obtained by dissolving lead powder with 5% nitric acid. For the Cu^{2+} solution, blue rhombohedral crystals of anhydrous copper nitrate $Cu(NO_3)_2 \cdot 6H_2O$ were dissolved in deionized water. Similarly, the Cd^{2+} solution was prepared by dissolving white prismatic crystal of cadmium nitrate tetrahydrate $Cd(NO_3)_2 \cdot 4H_2O$ in deionized water.

The concentrations of Pb^{2+} and Cd^{2+} in solution were determined using a graphite furnace atomic absorption spectrophotometer. Upon completion of the adsorption reaction, the absorbance of Pb^{2+} and Cd^{2+} was measured, and by referencing their respective calibration curves, the corresponding concentrations were derived through conversion. As for Cu^{2+} concentration, it was obtained indirectly: an ultraviolet spectrophotometer

FIGURE 5.1 Microstructures of the silty soil loaded with (a) raw sample, (b) Pb^{2+}, (c) Cu^{2+}, and (d) Cd^{2+}.

was employed to measure the complex formed between the solution and copper reagent (DDTC-Na), which presents as a thin white crystalline film. This complex is a yellow-brown colloid and remains stable for 30 min, thereby facilitating reliable measurement.

5.2.1.3 Test Apparatus

The one-dimensional soil column apparatus (Figure 5.2) is comprised of a test cylinder, a thermostatic control unit, a data acquisition system, and a water supply system (Bai et al., 2024). The test cylinder measured 60 cm in length, with an inner diameter of 14.8 cm and a wall thickness of 0.6 cm.

The test cylinder is constructed from two high-temperature-resistant acrylic cylinders, each 30 cm in length, which are joined via flanges and bolts. Both ends are similarly connected to end covers using flanges and bolts. This setup maintains a constant water-head for the soil column. Additionally, two rows of openings at equal height, spaced 7.5 cm apart vertically, were prepared along the cylinder's sidewall to facilitate the placement of moisture sensors, temperature sensors, and the collection of contaminated soil samples. Both the thermostatic control system and the water supply system incorporate a water tank and heat-resistant rubber hoses. A constant temperature was maintained by means of the water tank (temperature control range: 20–100 °C), with a peristaltic pump circulating water from the tank through hoses wrapped around the cylinder's outer wall, thereby achieving a stable ambient temperature. In terms of water supply, the tank provided seepage water at a

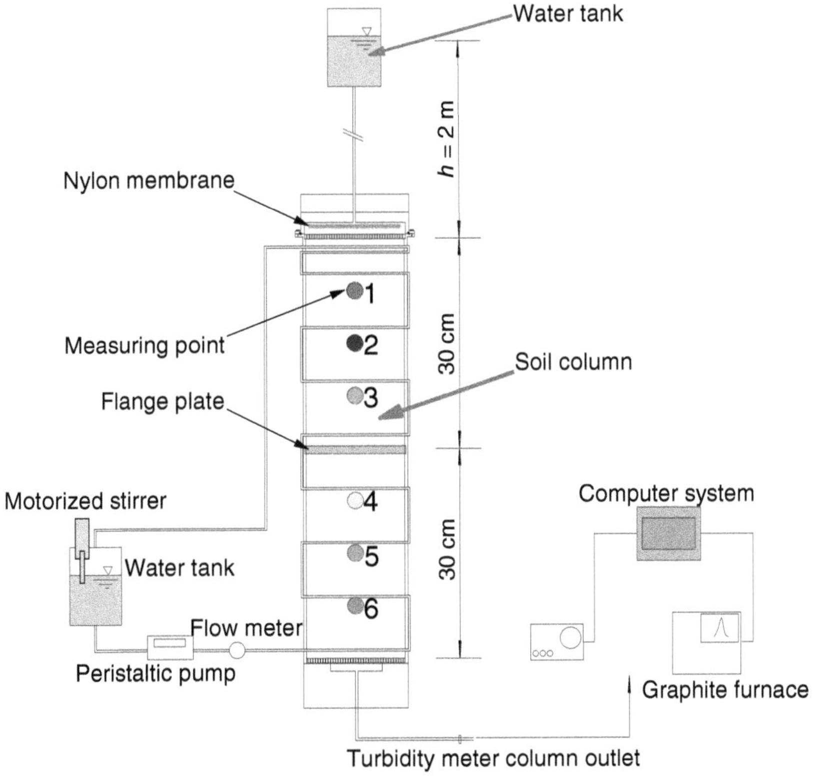

FIGURE 5.2 Experimental apparatus.

controlled, constant temperature. The data acquisition system comprises sensors and data loggers for measuring temperature and humidity. The platinum resistance temperature sensor offers high sensitivity, stability, and strong oxidation resistance. Its probe measures 2 mm in diameter and 15 mm in length. For moisture measurement, a soil moisture sensor with an accuracy of 0.1% was employed, alongside a dedicated soil water collector.

The readings from moisture sensors are significantly influenced by temperature variations. To calibrate the volumetric moisture content measurements, a dedicated small-scale sealed soil column sample (ensuring the same moisture content and dry density) was utilized. This calibration procedure involved adjusting the temperature of the soil specimen (temperature range: $T = 20 - 60°\mathrm{C}$) while simultaneously recording the moisture sensor readings. Test results revealed a linear fitting relationship between the sensor readings and temperature. Specifically, if the temperature at State 1 is T_1 with a corresponding sensor reading of θ_1, and at State 2 is T_2 with a reading of θ_2, the actual change in volumetric moisture content during this interval is given by $d\theta = (\theta_2 - \theta_1) - 0.0012 \times (T_2 - T_1)$. Regarding the adsorption tests of HMs onto the silty soil, the solution was first prepared and then agitated using a double-layer oscillator for 1 h (110 times/min). This ensured that the HMs fully reacted with soil particles and reached a stable adsorption state prior to measurement.

5.2.2 Attachment–Detachment Results of the Silty Soil

5.2.2.1 Test Method

The silty soil was oven-dried at 105 °C for 12 h until completely dry. Subsequently, multiple 0.2-g soil samples were measured using disposable weighing paper and placed into dry, clean conical flasks containing solutions of Pb^{2+}, Cu^{2+}, and Cd^{2+} at varying concentrations. These samples were then agitated on a mechanical shaker for 1 h (110 times/min). Based on prior adsorption-kinetics tests, the adsorption period was established as 24 h. Following the attachment–detachment reaction, the pollutant solution concentration for each specimen set was measured in triplicate, with the average value adopted as the final result to ensure reliability. The solution concentration range was set to 0–1200 μg/ml for adsorption, and 80–800 μg/ml for desorption, under three different temperatures (i.e., 20 °C, 40 °C, and 60 °C). For the desorption process at a specific concentration (e.g., 800 μg/ml), adsorbed HMs were gradually released by adding deionized water (i.e., by decreasing the concentration or by implementing a dilution process) until the solution concentration approached zero.

The soil column was prepared with an initial volumetric moisture content of 15.8% and a dry density of 1.3 g/cm^3. To achieve uniform moisture distribution, the silty soil filling was placed in 12 successive layers along the height of the test cylinder (5 cm per layer). After each layer was placed and compacted to the target height, its surface was scraped with a blade before the next layer was added. Once the column was completed, the end covers were attached to the cylinder flanges using bolts. The moisture/temperature sensor positions were labeled 1–6 from top to bottom (Figure 5.2). After sensor installation, the access holes were sealed with glass adhesive to prevent water leakage during seepage tests.

The soil column had an initial temperature of 20 °C. When required, the column temperature could be elevated via the thermostatic control device to achieve target values (e.g.,

$T = 40$ and 60 °C). Based on the concentration of the injected solution, seepage transport tests were conducted under five conditions: deionized water, Pb^{2+} solution (20 and 100 μg/ml), Cu^{2+} solution (100 μg/ml), and Cd^{2+} solution (100 μg/ml). All tests were performed with the soil column in a vertically orientation and under a constant water-head of $h = 2$ m to maintain one-dimensional seepage conditions. Naturally, the role of gravity cannot be disregarded in the analysis of the underlying mechanisms.

Upon commencement of the seepage tests, measurements of temperature, moisture content, and HMs concentration in the filtrate were conducted simultaneously. Following test completion, samples were collected from seven designated locations within the soil column (measuring points 1, 2, 3, 4, 5, and 6 and the infiltration location). At each sampling location, three replicate determinations were performed, and the average of these triplicate measurements was adopted as the final results.

HMs concentrations in the soil samples were quantified through the following procedure. First, 100 ml of a 0.15 mol/L EDTA solution was added to a conical flask containing 3 g of dried soil. The mixture was oscillated for 24 h to ensure complete chelation with adsorbed Pb^{2+}, Cu^{2+}, and Cd^{2+}, thereby extracting the HMs from the soil matrix. The resulting suspension was transferred to a centrifuge tube and centrifuged for 5 min to obtain a clear supernatant. Subsequently, the attachment quantity was measured based on this extracted solution.

5.2.2.2 Lead Ions (Pb^{2+})

The surface of soil particles exhibits numerous protrusions and indentations, contributing to a substantial specific surface area. This characteristic facilitates the binding of the silty soil with HMs in solution, leading to high adsorption efficiency. In practice, the adsorption of Pb^{2+} proceeded more rapidly at lower solution concentrations, with an efficiency exceeding 90%. As the Pb^{2+} concentration rose, the adsorption rate declined and the attachment quantity gradually approached its maximum value. For instance, the adsorption efficiency for Pb^{2+} dropped from 90.56% at 12.511 μg/ml to 32.97% at 1200 μg/ml.

From Figure 5.3, the characteristics of the Pb^{2+} attachment–detachment curves (see the theory model in Chapter 3; that is, Bai model; Bai et al., 2021b) remained consistent across different temperatures (20 °C, 40 °C, and 60 °C; see Figure 5.3a, b, and c, respectively). The adsorption rate began to decrease with increasing solution concentration, while the adsorbed quantity approached a constant value. The desorption rate was initially slow but tended to accelerate (e.g., 150 μg/ml in Figure 5.3b) once the Pb^{2+} concentration fell below a certain threshold. This behavior occurs because during desorption in the silty soil, the residual Pb^{2+} in the solution still exceeds that on the solid particles, although there is more Pb^{2+} at the high concentration adsorption point. At this time, the solubility product equilibrium constant is far from being reached. Consequently, the Pb^{2+} adsorbed onto the silty soil is not easily released, only the physically adsorbed fraction desorbs initially. As the dissolved Pb^{2+} concentration continues to decrease, the amount retained on the solid particles gradually becomes relatively greater, prompting more adsorbed Pb^{2+} to be released from the silty soil and resulting in faster desorption efficiency.

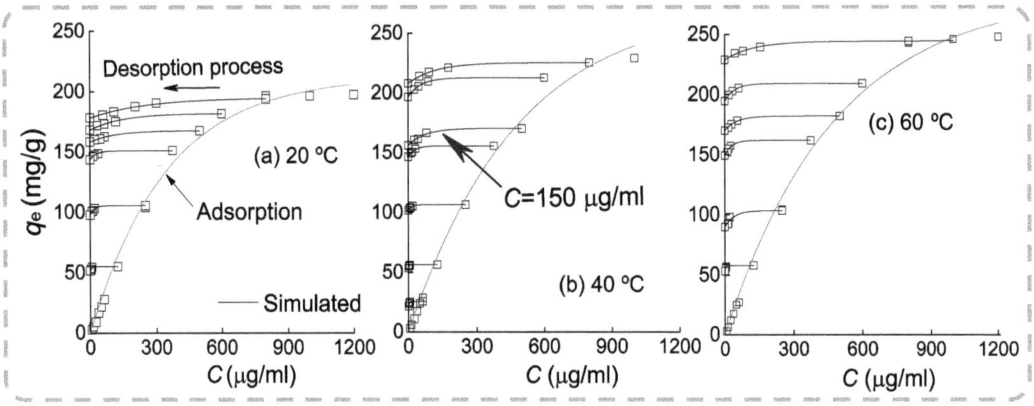

FIGURE 5.3 Attachment–detachment curves. (a) 20 °C. (b) 40 °C. (c) 60 °C.

The attachment quantity (q_e) was obtained using the following equation:

$$q_e = \frac{(C_0 - C_e)V}{m} \tag{5.1}$$

where V is the volume of solution, m is the mass of soil, C_0 is the initial concentration, and C_e is the residual concentration after a reaction occurs.

The relationship between the concentration C and the attachment quantity q_e can be described by the Bai model (see Chapter 3). Figure 5.3 gives the corresponding attachment–detachment process, which accounts for the so-called attachment-history effect. Static attachment–detachment parameters demonstrate that the proposed model effectively describes Pb^{2+} behavior on the silty soil, with a determination coefficient $R^2 > 0.97$. A further comparison was made among the attachment–detachment curves at different temperatures shown in Figure 5.3 (e.g., the initial Pb^{2+} concentration is 800 μg/ml). This comparison revealed that the adsorption amount of Pb^{2+} rose progressively with temperature, recording values of 193.98, 225.560, and 242.86 mg/g for temperatures of 20 °C, 40 °C, and 60 °C, respectively; adsorption efficiency likewise increased at the same concentration. Correspondingly, the adsorption equilibrium constant k_d exhibited an upward trend (i.e., 0.5694, 0.5746, and 0.5925 ml/mg at 20, 40, and 60 °C, respectively), while the desorption equilibrium constant κ_r gradually declined to 0.4735 ml/mg, 0.4601 ml/mg, and 0.4338 ml/mg, respectively.

The observed phenomenon can be explained by the fact that elevated temperature enhances the kinetic energy of Pb^{2+} ions, altering their activation energy toward the transition state and thereby facilitating their adsorption onto the silty soil. During the desorption process (Figure 5.3), the temperature rises similarly enhanced the residual attachment of Pb^{2+}, indicating an overall improvement in the soil's affinity for Pb^{2+}. In practice, higher temperature intensifies water turbulence within the soil pores, and the thermodynamic driving force is amplified due to the stochastic Brownian motion of HMs. Elevated temperature raises the collision probability between soil particles and HMs, leading to more frequent and effective interactions, which in turn strengthen the overall adsorption effect.

5.2.2.3 *Copper Ions* (Cu^{2+})

Experimental results indicated that the maximum attachment quantity for Cu^{2+} was notably lower than that for Pb^{2+}, a behavior distinctly different from the attachment–detachment of Cu^{2+} on quartz sand. Increasing the concentration of Cu^{2+} also elevated its attachment quantity on the silty soil. Furthermore, the tests revealed that within a Cu^{2+} concentration range of 10 μg/ml to 60 μg/ml, the attachment quantity varied from 4.792 mg/g to 39.184 mg/g, accompanied by an adsorption efficiency exceeding 90%. As the Cu^{2+} concentration continued to rise, the adsorption efficiency began to decline, which is similar to the case observed for Pb^{2+}. The attenuation factors for the adsorption and desorption processes were $\beta_1 = 0.0066$ $(R^2 = 0.99)$ and $\beta_2 = 0.0940$ $(R^2 = 0.95)$, respectively. The corresponding attenuation coefficients for Pb^{2+} were $\beta_1 = 0.0027$ $(R^2 = 0.99)$ and $\beta_2 = 0.0714$ $(R^2 = 0.95)$, indicating that the nonlinear behaviors of Cu^{2+} were more pronounced than that of Pb^{2+}. The adsorption of Cu^{2+} onto quartz sand is primarily attributed to the hydrolysis of Cu^{2+}, with the resulting copper hydroxide precipitate adhering to the sand surface via dispersion. Moreover, Cu^{2+} desorption in that system is driven mainly by hydrodynamic flushing, exhibiting predominantly linear characteristics. In contrast, the silty soil comprises multiple mineral components and possesses a larger specific surface area. Consequently, the amount of Cu^{2+} adsorbed onto the silty soil is significantly greater than onto quartz sand. Beyond physical adsorption, notable chemical adsorption also occurs. As a result, Cu^{2+} bound to the silty soil is difficult to remove during desorption, leading to the observed pronounced nonlinear characteristics.

Similar to Pb^{2+}, the quantity of the adsorbed Cu^{2+} rose at higher temperatures during the adsorption stage, following a consistent trend in the adsorption equilibrium constant. In contrast, the desorption equilibrium constant exhibited the opposite behavior. This difference stems from the fact that Cu^{2+} adsorption is an endothermic process, while its desorption is exothermic; consequently, a temperature increase tends to inhibit the desorption reaction of Cu^{2+}.

5.2.2.4 *Cadmium Ions* (Cd^{2+})

The attachment–detachment behavior of Cd^{2+} showed similarities to that of Pb^{2+} and Cu^{2+}. Likewise, a temperature rise progressively enhanced the adsorption quantity of Cd^{2+}. The desorption equilibrium constant κ_r for Cd^{2+} decreased from 0.1055 ml/mg to 0.0737 ml/mg as the temperature increased from 20 °C to 60 °C, which is associated with the endothermic nature of Cd^{2+} adsorption. Nevertheless, the effect of temperature change on the desorption reaction is considerably less pronounced than on the adsorption process, resulting in a significantly higher residual attachment compared to that at ambient temperature.

The proposed attachment–detachment model (i.e., Bai model; see Chapter 3) advances traditional isothermal attachment–detachment frameworks. It effectively captures the influences of temperature, HMs types, and attachment–detachment history, as represented through equilibrium constants and attenuation coefficients.

5.2.3 Attachment–Detachment Mechanism

Among the tested HMs, Pb^{2+} exhibited the largest attachment quantity, followed by Cu^{2+} and then Cd^{2+}. At a solution concentration of 800 μg/ml and 20 °C, the maximum adsorption capacities reached 193.976, 88.854, and 75.672 mg/g, respectively. During the subsequent desorption process, the peak residual adsorption amounts correspondingly decreased to 178.363, 70.358, and 56.9571 mg/g, respectively.

To examine microstructural variations in the silty soil loaded with HMs, solutions of Pb^{2+}, Cu^{2+}, and Cd^{2+} were formulated at specific concentrations of 50 and 100 μg/ml. Each solution was combined with 5 g of the silty soil in a beaker and allowed to interact for 24 h. Subsequently, the silty soil samples were separated via vacuum filtration, dried, and prepared for SEM imaging.

Figure 5.1b–d reveals that the microstructures of the soil particle surface underwent significant alterations upon the adsorption of HMs. Specifically, the effect of Pb^{2+} caused the flaky structure (hydrolysate) on the surface of soil particles, the influence of Cd^{2+} resulted in a superposition of the flaky structure, while the effect of Cu^{2+} caused the flocculent phenomenon to be more serious, resulting in the reduction of the specific surface area of soil particles (compared with the SEM image of the raw sample; Figure 5.1a). Relevant studies indicate that the adsorption of Pb^{2+} and Cd^{2+} is predominantly an ionic reaction, while Cu^{2+} adsorption occurs mainly via ionic exchange. Consequently, the microstructure of the silty soil loaded with Cu^{2+} is distinctly different from that observed for Pb^{2+} and Cd^{2+}.

5.2.3.1 Lead Ions (Pb^{2+})

For the adsorption test, the Pb^{2+} solution with pH = 7 was employed. Following the experiment, the solution pH was observed to drop below 7. In the corresponding XRD diffraction pattern (Figure 5.4), the characteristic peaks of calcite in the silty soil diminished after Pb^{2+} adsorption. The EDS spectrum (Figure 5.5) further revealed a notable decrease in oxygen content, along with reductions in Ca and Mg levels, indicating that Pb^{2+} had reacted with a portion of the calcite present. The reaction process can be described by Eqs. (5.2)–(5.4):

$$H_2O + Pb^{2+} \rightarrow Pb(OH)^+ + H^+ \tag{5.2}$$

$$Pb(OH)^+ + CaCO_3 \rightarrow PbCO_3 + Ca^{2+} + OH^- \tag{5.3}$$

$$Pb(OH)^+ + CaMg(CO_3)_2 \rightarrow 2PbCO_3 + Ca^{2+} + 2OH^- + Mg^{2+} \tag{5.4}$$

The XRD mapping for compound analysis indicated the presence of a minor quantity of $Pb_2SO_4(OH)_2$ within the soil sample. Furthermore, EDS energy spectrum analysis showed

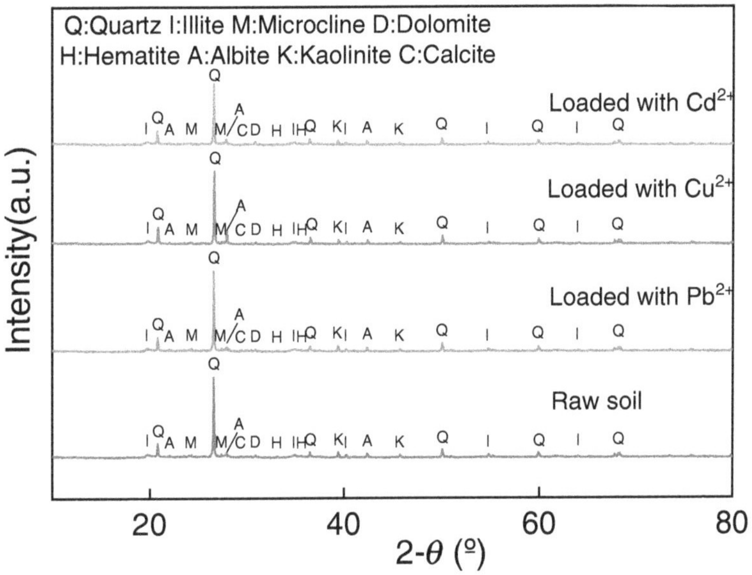

FIGURE 5.4 XRD infraction pattern loaded with (a) Pb^{2+}, (b) Cu^{2+}, and (c) Cd^{2+}.

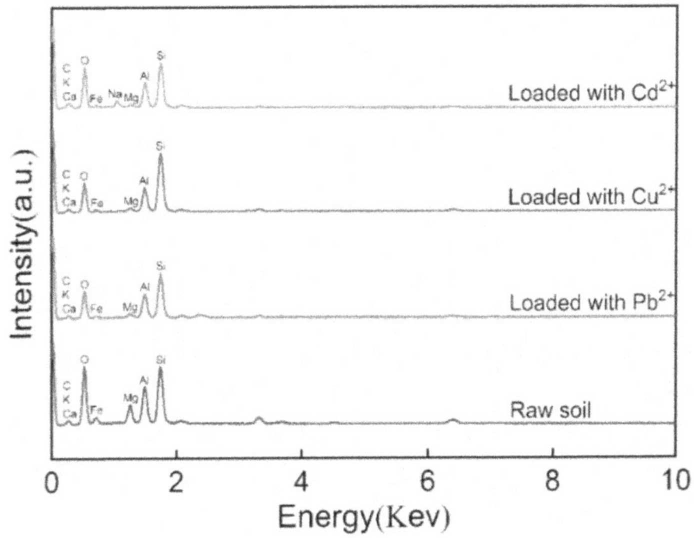

FIGURE 5.5 EDS spectrum loaded with (a) Pb^{2+}, (b) Cu^{2+}, and (c) Cd^{2+}.

the decreased relative contents of Mg, Fe, O, Al, and Si. Indeed, the anisotropic interlayer charges inherent in clay minerals can induce coordination reactions, as represented by Eqs. (5.5) and (5.6) or cation exchange (Eq. (5.6)). The XRD diffraction pattern additionally illustrated a reduction in the composition of minerals such as kaolinite and plagioclase, confirming that HMs are retained by clay minerals primarily in the form of complexes:

$$2 \equiv SOK + Pb^{2+} \rightarrow \equiv SPbOS + 2K^{2+} \tag{5.5}$$

$$2 \equiv SO^- + Pb^{2+} \rightarrow \equiv SOPbOS \tag{5.6}$$

where S represents the mineral composition and K represents the HMs held by clay minerals.

From Figure 5.6, in the FT-IR spectrum, the absorption peak at 1436 cm^{-1} vanished after the silty soil was loaded with Pb^{2+}, suggesting that the —COOH groups in kaolinite had combined with Pb^{2+}. Additionally, the peak at 430 cm^{-1} disappeared, which correlates with the reduced relative content of Si following Pb^{2+} loading. Figure 5.4 demonstrates that the corresponding XRD peaks at 68.82° and 59.94° and diminished after Pb^{2+} adsorption. Although the Si—O bond in quartz is relatively stable, its crystal surfaces may still undergo hydrolysis (Eq. (5.7), leading to precipitate formation:

$$\equiv \text{Si-O-Si} \equiv + Pb(OH)^+ + H_2O \rightarrow \equiv \text{Si-O-Pb}(OH) + \equiv Si(OH) + H^+ \tag{5.7}$$

5.2.3.2 *Copper Ions* (Cu^{2+})

From Figure 5.4, the XRD diffraction pattern of the silty soil after loading with Cu^{2+} reveals significantly enhanced diffraction peaks at 20.88°, 27.94°, and 59.96°, corresponding to illite, albite, and kaolinite, respectively. Furthermore, new diffraction peaks emerged at 12.32° and 41.14°, indicating that the mineral composition of the silty soil plays a key role in Cu^{2+} adsorption. A decrease in the calcite diffraction peak at 65.32° was observed, demonstrating that Cu^{2+} reacted with a portion of the calcite present. The pH of the solution after adsorption was measured 6.85, which is below 8. Based on the solubility product equilibrium, the precipitation equilibrium constant was not reached, $Cu(OH)_2$ could not form. The reduction in the calcite peak intensity in the XRD pattern, along with a decreased oxygen content, is consistent with the process described by Eq. (5.8):

$$CaCO_3 + Cu^{2+} \rightarrow CuCO_3 + Ca^{2+} \tag{5.8}$$

Figure 5.5 indicates a reduction in Mg content, signifying that the adsorption reaction on illite and kaolinite forms new complexes. The adsorption mechanism of Cu^{2+} is described by Eqs. (5.9) and (5.10):

$$\equiv SOH + Cu^{2+} \rightleftharpoons \equiv SOCu^{2+} + H^+ \tag{5.9}$$

$$\equiv SOH + Cu^{2+} + H_2O \rightleftharpoons \equiv SOCuOH + 2H^+ \tag{5.10}$$

In the untreated silty soil, weak adsorption peaks were observed at 2362 cm^{-1} (Figure 5.6). Following the adsorption of Cu^{2+}, these peaks disappeared. Additionally, the absorption peak at 1436 cm^{-1}, present in the raw sample and associated with carboxyl

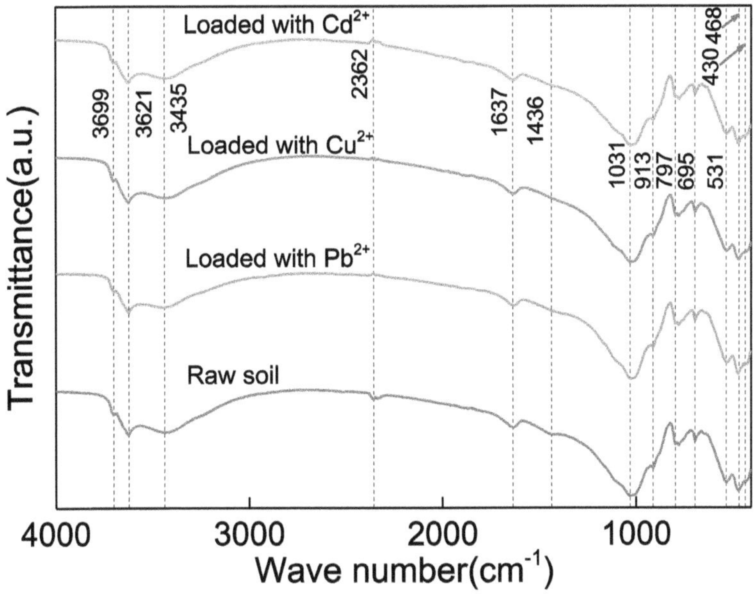

FIGURE 5.6 FT-IR functional group loaded with (a) Pb^{2+}, (b) Cu^{2+}, and (c) Cd^{2+}.

functional groups, also vanished after Cu^{2+} adsorption. The interaction between Cu^{2+} and carboxyl groups can be expressed by

$$-COOH + Cu^{2+} \rightarrow -COOCu^{+} + H^{+} \tag{5.11}$$

5.2.3.3 Cadmium Ions (Cd^{2+})

For the case of Cd^{2+}, the transmittance peaks at 3699 and 3621 cm^{-1} corresponding to —OH functional groups present in the clay minerals of the silty soil (Figure 5.6). Peaks at 1031, 797, and 695 cm^{-1} were attributed to Si—O functional groups, while those at 531 and 468 cm^{-1} represented Si(Al) groups. These peaks remained unchanged after Cd^{2+} adsorption, indicating that these functional groups did not react with Cd^{2+}. This observation is consistent with the absence of a notable decrease in Si content shown in Figure 5.5. The transmissivity at 3435 and 1637 cm^{-1} resulted from the stretching vibration of clay mineral molecules following water adsorption. After loading with Cd^{2+}, these wave numbers did not shift significantly, suggesting that the associated functional groups were not involved in reactions with Cd^{2+}. These signals are likely related to both bound water in the silty soil and moisture adsorbed from the atmosphere. It is particularly noteworthy that the absorption peak at 1436 cm^{-1} in the silty soil disappeared after Cd^{2+} adsorption, which is associated with the bending vibration of the carboxyl functional group (—COOH) in the mineral:

$$-COOH + Cd^{2+} \rightarrow -COOCd^{+} + H^{+} \tag{5.12}$$

Compared to the raw sample, the intensities of the diffraction peaks at 24.24°, 29.4°, 24.22°, 26.729°, 27.72°, 28.22°, 29.4°, and 50.1° diminished following the adsorption of Cd^{2+} (Figure 5.4). Notably, the enhanced peak at 27.88° was associated with the presence of albite, kaolinite, and illite. Kaolinite possesses a point of zero charge around 5 and tends to adsorb Cd^{2+} at pH >4, the measured pH after Cd^{2+} adsorption was 6.8. Moreover, Figure 5.5 reveals reductions in the contents of Si , Al , and O . Consequently, Cd^{2+} reacts with silicate minerals to form complexes of $CdSiO_3$ and $CdAl(SiO_4)_2$. This reaction process can be expressed as follows:

$$S \equiv OH \rightleftharpoons = SO^- + H^+ \tag{5.13}$$

$$S \equiv O^- + Cd^{2+} \rightleftharpoons = CdSO^- \tag{5.14}$$

where S refers to the clay minerals in the silty soil.

The measured zeta potentials of the silty soil treated with Pb^{2+}, Cu^{2+}, and Cd^{2+} were 1.76, −9.64, and −11.3 mV, respectively, while the untreated silty soil exhibited a zeta potential of −7.78 mV. In most clay minerals, particles surfaces carry a net negative charge due to their inherent mineral compositions. Through ion-exchange, coordination, and complexation reactions, the silty soil surface becomes appreciably loaded with HMs. However, the specific interactions between different HMs and silty soil vary, leading to distinct final zeta potentials. Overall, the order of zeta potential is Pb^{2+}, Cu^{2+}, and Cd^{2+}, which is consistent with the order of adsorption capacities in the static attachment–detachment tests presented in Section 5.2.4.

These microscopic alterations confirm that the attachment–detachment behavior is closely associated with the type of chemical reaction (e.g., ion reaction, ion exchange) as well as changes in surface potential and functional group activity.

5.2.4 Transport of the Aqueous Solution Contaminated by HMs

5.2.4.1 Influence of HMs on the Movement of Moisture

Figures 5.7 and 5.8 illustrate the temporal variation of moisture content when deionized water, Pb^{2+} , Cu^{2+} , and Cd^{2+} solutions were introduced into an unsaturated silty soil column. The moisture content within the column increased and redistributed over time, transitioning particularly from an unsaturated to a saturated condition. In the deionized water experiment (Figure 5.7a), the peak volume moisture content progressively advanced to locations further from the infiltration surface (i.e., measuring points 1, 2, 3, 4, 5, and 6 in Figure 5.2) after 140 h. For the case of injecting Pb^{2+} solution (Figure 5.7b and c), Cu^{2+} solution (Figure 5.8a), and Cd^{2+} solution (Figure 5.8b), there were similar evolution rules.

For a given HMs (e.g., Pb^{2+}), a higher concentration of the injected solution resulted in a faster arrival of the wetting front at the same measuring point. Taking Point 4 as an example (Figure 5.7), as the concentration increased (0→20 μg/ml →100 μg/ml), the time required for the wetting front to reach this point decreased from 25 h to 19 h and then 17 h. Correspondingly, the time to reach the moisture inflection point shortened from 38 h to 36 h and further to 28 h. Moreover, with increasing distance from the injection surface,

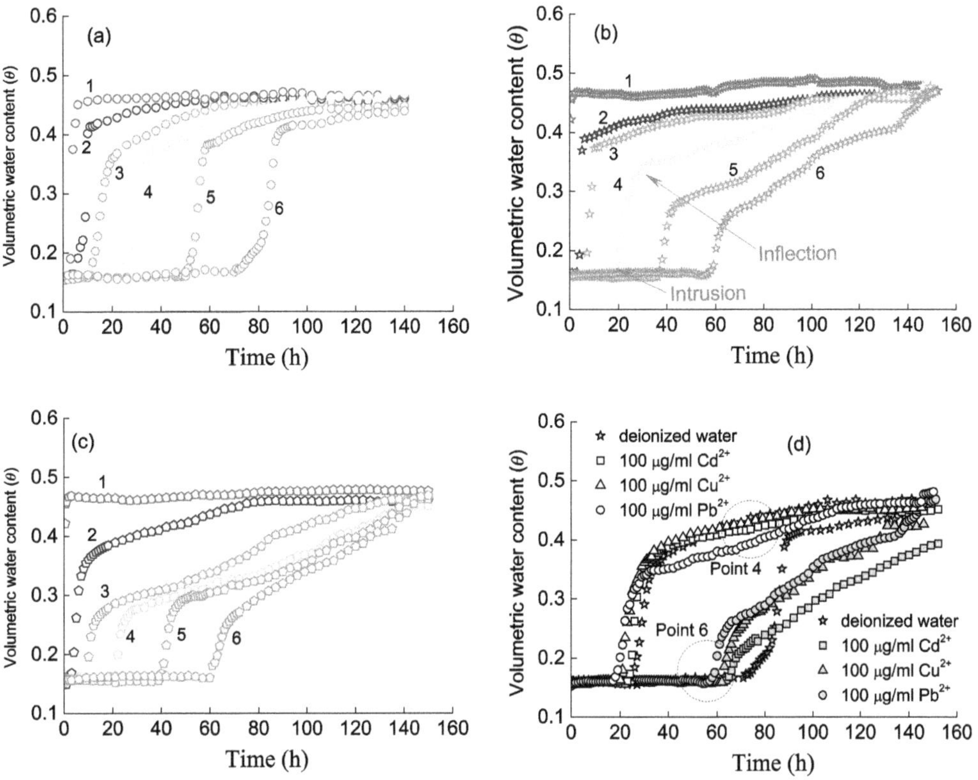

FIGURE 5.7 Moisture transport over time loaded with Pb^{2+}. (a) Deionized water. (b) With a concentration of 20 μg/ml. (c) With a concentration of 100 μg/ml. (d) Comparison of different distances.

the wetting front advanced more rapidly compared to the flow of deionized water. Additionally, the moisture inflection point occurred earlier and at a lower volumetric moisture content as the distance increased.

The surface properties of soil particles change when loaded with HMs. Specifically, the interaction of HMs with mineral components elevates the solute potential in the pore water and reduces the water-retention capacity of soil particles. Consequently, during water infiltration, moisture moves more readily from zones of lower matric suction to those with higher suction. In other words, preferential flow develops during HMs transport, enabling the wetting front to reach monitoring points in a shorter duration.

Comparative experiments using identical concentrations but varying types of HMs (Figures 5.7d and 5.8; 100 μg/ml solution) demonstrated that Pb^{2+} exhibited the highest mobility, followed by Cu^{2+}, while Cd^{2+} showed the lowest transport rate, as observed at characteristic locations (e.g., points 4 and 6).

5.2.4.2 Transport Process of HMs

At an injection concentration of 100 μg/ml for HMs, the effluent concentration remained below 100 ng/ml (Figure 5.9), indicating that the silty soil adsorbed over 90% of the HMs.

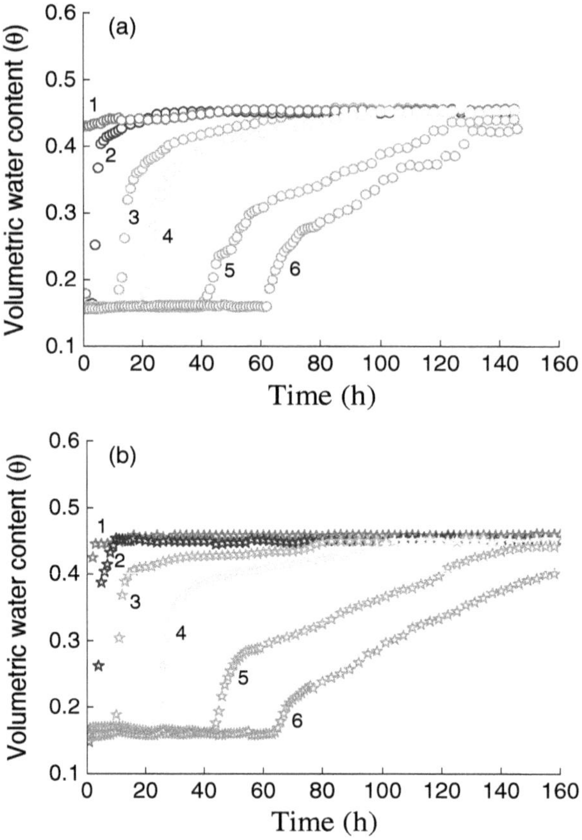

FIGURE 5.8 Moisture transport over time. (a) Cu^{2+}. (b) Cd^{2+}.

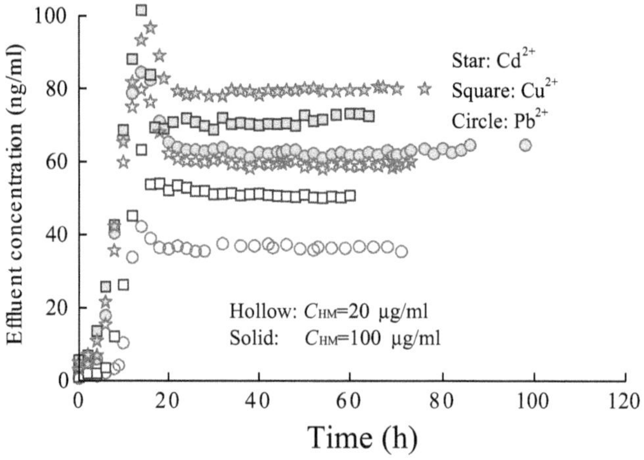

FIGURE 5.9 Transport processes at different injected concentrations.

Notably, a higher adsorbed amount of HMs (e.g., Pb^{2+} compared with Cd^{2+}) corresponded to a shorter breakthrough time and a lower peak concentration of the effluent. For a given injection concentration (e.g., $C_{HM} = 100\ \mu g/ml$), greater HMs adsorption more significantly alters the surface properties of soil particles, thereby reducing their water-holding capacity and matrix suction. This promotes preferential flow and leads to earlier breakthrough. It should be noted that the HMs concentrations used in these experiments substantially exceed typical levels in contaminated field soils, which underscores the potential of the silty soil as an effective, long-term adsorbent material.

Upon the arrival of the wetting front at the column base and the transition of the soil from an unsaturated to a saturated state, HMs primarily migrated along pathways characterized by reduced matrix suction, a condition influenced by the presence of the HMs themselves. Consequently, the HMs concentration in the effluent initially rose during the early process (e.g., in the time range of 0–15 h), reaching a peak value. However, as the soil progressively saturated, HMs percolated through the internal pore network of the porous medium. Subsequently, the outflow concentration declined rapidly before stabilizing at a steady state. This transition marks a critical turning point governing HMs transport behavior. Experimental findings demonstrate that the presence of HMs modifies the interface properties of soil particles (Figures 5.4–5.6), which ultimately influences the transport dynamics of the filtrate. Moreover, HMs significantly alter the matrix suction of unsaturated soil, a key factor governing water movement. The role of temperature is also considerable (Figure 5.3). Quantifying the matrix suction of soils laden with different HMs under varying temperatures is therefore essential for developing accurate transport models, representing a pertinent direction for future research.

Additionally, the effluent concentration of HMs increased with higher injection concentration (20 μg/ml→100 μg/ml), and the peak concentration of the corresponding breakthrough curve was also elevated. In fact, based on the earlier static attachment–detachment test, at an injection concentration of 100 μg/ml, the attachment quantity per unit mass of soil remained well below its saturation point.

5.2.4.3 Temperature Influence on Transport Processes

Figure 5.10 presents moisture movement patterns at two characteristic locations (points 4 and 6) within the soil column when the injected water contained HMs (loaded with Pb^{2+} and Cd^{2+}; see Figure 5.10a and b). From Figure 5.10, rising temperature (i.e., 20–60 °C) increases the kinetic energy of water molecules and promotes the transition of soil water from a bound state to free water, reflecting a distinct thermodynamic driving force. Consequently, the moisture transport is markedly accelerated, leading to a noticeable reduction in soil moisture content at any given time under elevated temperatures.

Elevated temperature enhances the adsorption affinity between HMs and soil particles (see Figures 5.1 and 5.3), which consequently reduces the concentration of HMs (e.g., Pb^{2+} and Cd^{2+}; Figure 5.11). For an injection concentration of 100 μg/ml, the difference in effluent concentration between 20 °C and 60 °C ranges approximately between 10% and 15%,

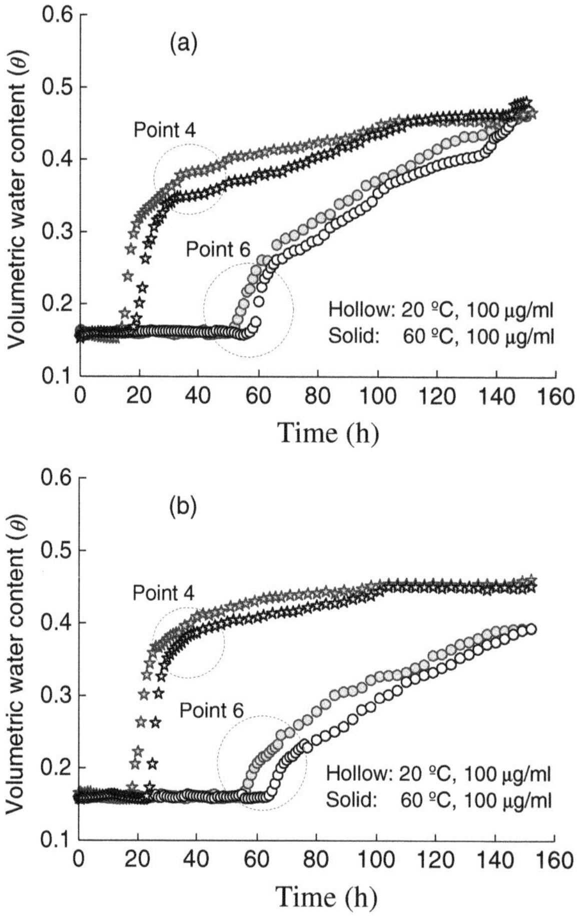

FIGURE 5.10 Temperature effect on the moisture transport loaded with (a) Pb^{2+} and (b) Cd^{2+}.

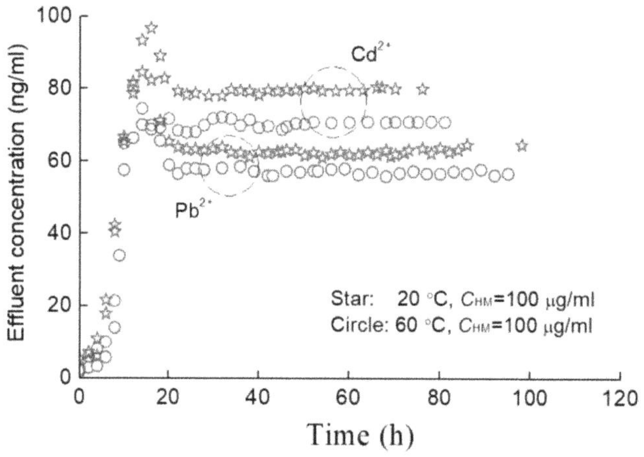

FIGURE 5.11 Temperature influence on the HMs transport.

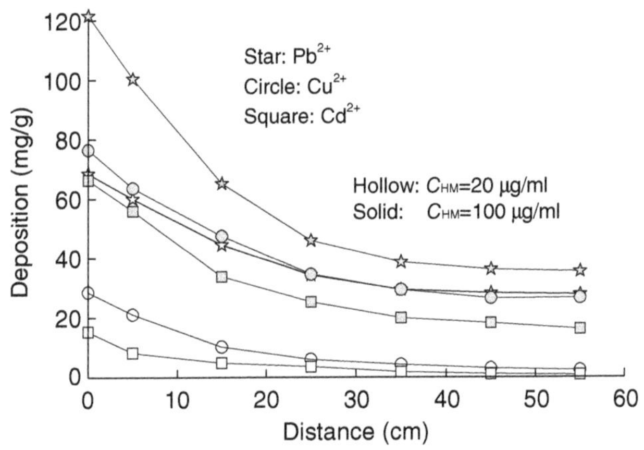

FIGURE 5.12 Concentration of HMs along the transport distance.

underscoring the significant influence of temperature. Owing to its stronger adsorption affinity with soil particles compared to Cd^{2+}, the Pb^{2+} consistently exhibits lower effluent concentrations under identical conditions.

5.2.4.4 Distribution of HMs along the Soil Column

Figure 5.12 illustrates that the transport behaviors of different HMs in unsaturated soil exhibit distinct variations. In general, both the transport distance and retained concentration of Pb^{2+} exceed those of Cu^{2+} and Cd^{2+}, aligning with the affinity sequence in the static attachment–detachment tests. Higher HMs adsorption corresponds to more pronounced alteration of the soil microstructure, fostering preferential flow paths during seepage. Consequently, the HMs concentrations become markedly heterogeneous along the soil column, diminishing with increasing distance from the injection surface. While Cu^{2+} and Cd^{2+} display spatial distributions similar to Pb^{2+}, their overall adsorption capacities are comparatively lower. Moreover, a higher injection concentration consistently leads to significantly greater HMs retention within the soil column than a lower concentration does.

Investigations into contaminant transport (e.g., HMs) through unsaturated silty soil, employed as a porous medium, can finish foundational insights and practical guidance for the future remediation of subsurface contamination.

5.3 EFFECTIVE THERMAL CONDUCTIVITY

5.3.1 Simulation on Effective Thermal Conductivity–SPH Method

The SPH method is a Lagrangian, mesh-free particle algorithm. In this approach, the computational domain is discrete set of particles. The discretization of a field function involves two key stages. First, the value of a field function $f(\mathbf{x})$ at a point $\mathbf{x}$ is expressed in integral form as follows:

$$f(\mathbf{x}) \approx \tilde{f}(\mathbf{x}) = \int_{\Omega} f(\mathbf{x}')W(|\mathbf{x}-\mathbf{x}'|, h)d\mathbf{x}' \tag{5.15}$$

where Ω, W, and h denote the integral domain, the kernel function, and the smoothing length of the kernel function, respectively.

A frequently used kernel function, the cubic *B*-spline function, is adopted (Bai et al., 2018):

$$W(R,h)=\alpha_{\mathrm{d}}\times\begin{cases}2/3-R^2+R^3/2, & 0\le R<1\\ (2-R)^3/6, & 1\le R<2\\ 0, & R\ge 2\end{cases} \tag{5.16}$$

Here R is the relative distance and is equal to r/h, where r is the particle distance. In 1D, 2D, and 3D spaces, the normalized values of the constant α_{d} are $1/h$, $15/7\pi h^2$, and $3/2\pi h^3$, respectively.

Next, the integral form of the field function is discretized into a series summation. The field function at particle *i* is approximated as

$$f(\mathbf{x}_i)\approx\sum_{j=1}^{N}\frac{m_j}{\rho_j}f(\mathbf{x}_j)W(|\mathbf{x}_i-\mathbf{x}_j|,h) \tag{5.17}$$

where $\mathbf{x}$ is the distance vector; m and ρ are the mass and density of particle j, respectively; the subscript j denotes a particle within the influence area of particle i, and N is the total number of particles.

Nevertheless, the conventional SPH algorithm presents inherent limitations when addressing complex engineering challenges. To overcome these shortcomings, numerous researchers have developed enhanced computational techniques, providing significant impetus for the continued advancement of the method. Concurrently, the application scope of SPH is expanding, now encompassing areas such as pore-scale heat conduction and the analysis of thermal consolidation in geotechnical materials (Rao and Bai, 2020; Bai et al., 2022b).

The differential equation for unsteady heat conduction within a three-dimensional isotropic medium without an internal heat source is

$$\frac{\partial T}{\partial t}=\frac{\lambda}{\rho c}\cdot\nabla^2 T \tag{5.18}$$

where λ is the effective thermal conductivity, ρ is the density, c is the specific heat, and T is the temperature.

The SPH approximation at point i can be written as follows:

$$\left(\frac{\partial T}{\partial t}\right)_i=\frac{1}{\rho_i c_i}\sum_{j=1}^{N}\frac{m_j}{\rho_j}\cdot\frac{4\lambda_i\lambda_j}{\lambda_i+\lambda_j}\left(T_i-T_j\right)\cdot\frac{\mathbf{x}_i-\mathbf{x}_j}{r_{ij}^{\,2}}\nabla_i W_{ij} \tag{5.19}$$

where $r_{ij}=|\mathbf{x}_{ij}|$.

The size of each time interval is suggested to be

$$\Delta t \leq \min\left(\frac{0.15h^2}{\kappa}\right) \tag{5.20}$$

where κ is the thermal diffusion coefficient, which is equal to $\lambda / \rho c$.

The explicit Euler method is employed for the integration step, $\tilde{f}(t+\Delta t) = f(t) + \Delta t \left.\frac{\mathrm{d}f}{\mathrm{d}t}\right|_t$. SPH has different discretization schemes for different derivation processes. Among these schemes, Eq. (5.19) is the most suitable for modeling thermal conduction. Its accuracy in 1D, 2D, and 3D spaces has been verified by extensive numerical tests.

The particle flow code (PFC) method is utilized to generate a 3D granular medium (Bai et al., 2022b). Originating from the discrete element method, the PFC approach was pioneered by Cundall and Strack (1979), defining discrete elements as rigid spheres. Using PFC3D 3.10 software (developed by ITASCA Inc., USA), the creation of a 3D dry granular assembly involves five primary stages. Initially, the software establishes rigid walls to define the specimen boundaries. Subsequently, the number and size distribution of spheres are specified, and these spheres are randomly generated within the computational domain. Third, gravity is applied, allowing the spheres to settle freely. Following natural accumulation, a moving plate is employed to compress the assembly slowly until a predefined porosity is achieved. Finally, the software exports the radius and centroid coordinates for each sphere, enabling the medium to be discretized into solid and gas particles, as illustrated in Figure 5.13a.

From Figure 5.13a, the cubic medium (named SAP1) has a side length of $L = 0.008$ m and consists of 70 spherical granules, each with an identical radius a = 1 mm. Following spatial discretization, the medium (Figure 5.13b) comprises a total of 64,000 particles, with a particle spacing of 0.0002 m. These particles are categorized as either gas or solid particles. The porosity is specified as 0.43. The calculation time is about 3 h.

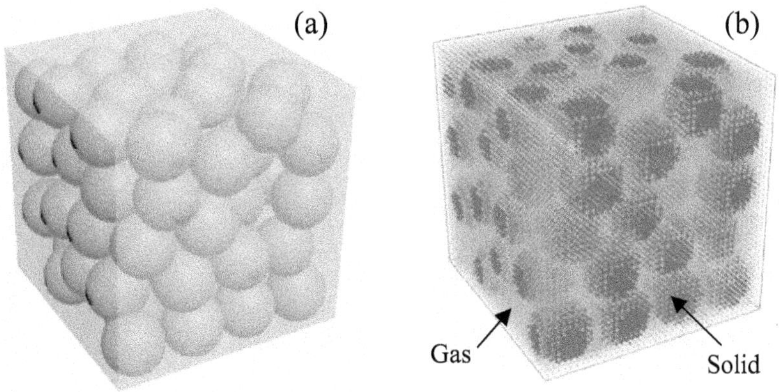

FIGURE 5.13 Solid and gas particles. (a) Restructured 3D porous medium. (b) Discretization in the dry state.

5.3.2 Humidifying Process

During the humidification of a porous medium from an initially dry condition, the evolution in the distributions of pore water and air phases can be approximately described. Initially, liquid bridges emerge within constricted pore spaces, water fills certain dead-end pores, and water films develop on the surfaces of solid granules. As water saturation increases, these liquid bridges expand and the water films thickness will also increase. Subsequently, adjacent liquid bridges gradually interconnect. When the air fraction further decreases, the continuity between air bubbles is disrupted, eventually leaving only isolated bubbles trapped within the pore network.

Figure 5.14 presents an illustrative 2D CT scanning image, where the humidification process is approximated in line with the aforementioned. This process can be broadly divided into two stages. First, gas particles are ranked based on their degree of interaction with the solid surface. Subsequently, these gas particles are sequentially converted into water particles. The initial step employs the particle pairing algorithm within the SPH method. In the dry state, the detection zone of a gas particle situated near a solid surface contains both solid and gas particles (Figure 5.15). A parameter p is defined as the proportion of solid particles within this detection zone. As illustrated in Figure 5.15, the p value at point A is greater than that at point B, and the p value for point B is greater than that for point C. According to their evolution, these particles located in dead-end pores (e.g., particle A) or near the surface of the solid phase (e.g., particle B) are wetted before those located in the centers of the pore channels (e.g., particle C). Moreover, the p values of the former are higher than those of the latter. Thus, the wetting sequence of gas particles generally follows the descending order of their p values.

From Figure 5.15, the identification zone resembles the influence domain of a particle, with its radius defined as a multiple of the smoothing length. Accordingly, the humidification process for SAP1 can be simulated using this algorithm (Figure 5.16).

The steady-state method represents a conventional laboratory approach for determining thermal conductivity. Under equilibrium conditions, the rate of heat transfer equals the

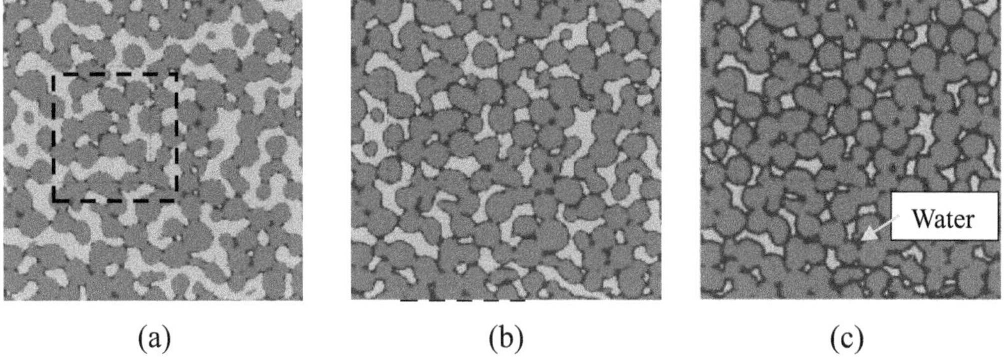

FIGURE 5.14 Humidifying process. (a) $S_r = 0.24$. (b) $S_r = 0.53$. (c) $S_r = 0.83$.

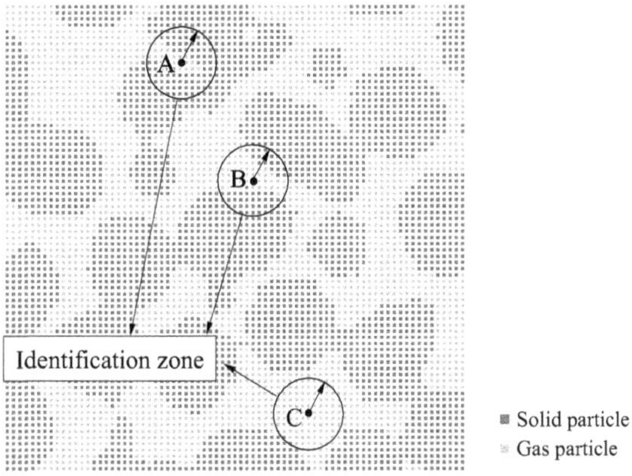

FIGURE 5.15 Identification zone due to the influence domain of a particle.

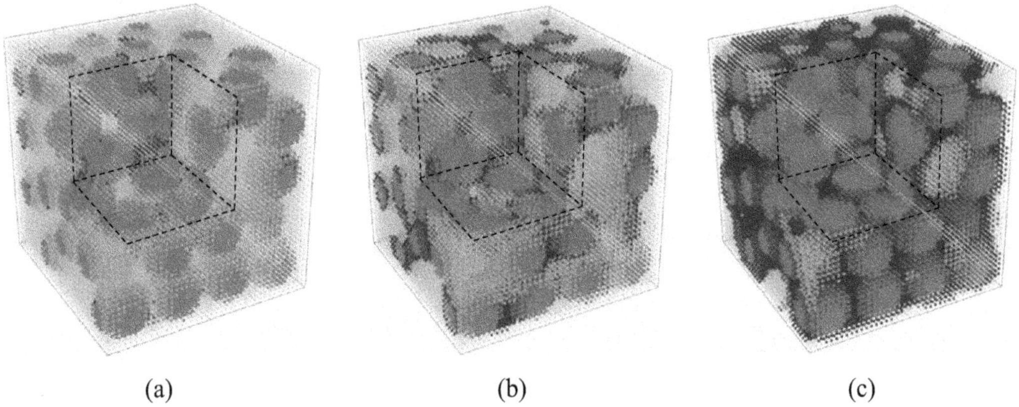

FIGURE 5.16 Simulation of the humidifying process for a 3D granular medium. (a) $S_r = 0.12$. (b) $S_r = 0.40$. (c) $S_r = 0.84$.

rate of heat dissipation in a stable conduction process. For an ideal infinite plate, the heat flux q satisfies the relationship:

$$q = \lambda \frac{\Delta T}{L} \tag{5.21}$$

where ΔT is the temperature difference between the two sides of the plate, and L is the thickness of the plate.

In the computational model, an infinite plate is simulated by imposing periodic boundary conditions on a square or cubic medium. The boundary conditions for the 2D simulation are illustrated in Figure 5.17. The top and bottom edges of the square medium are defined as periodic boundaries, allowing the medium to be periodically extended into an infinite plate of thickness L. The left and right boundaries are assigned different fixed

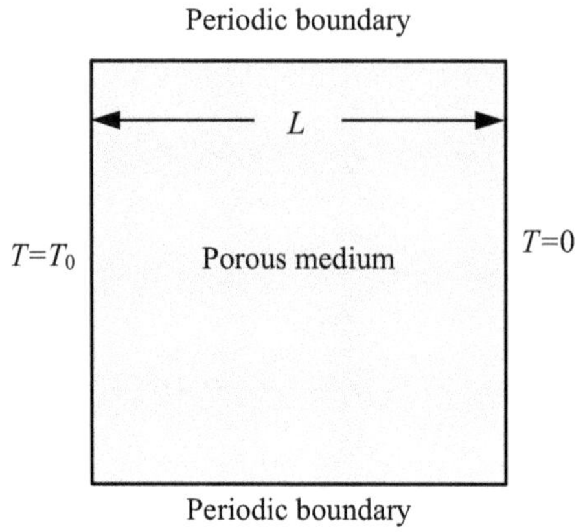

FIGURE 5.17 Boundary conditions of the steady-state simulation.

temperatures, establishing the initial thermal gradient. For the set of SPH particles, if the imposed temperature difference is T_0, the average heat flux is calculated as

$$\bar{q} = \frac{1}{N}\sum_{i=1}^{N}\lambda_i\left(\frac{\mathrm{d}T}{\mathrm{d}x}\right)_i \tag{5.22}$$

Here, if the temperature field reaches a steady state, the thermal conductivity of the medium is determined by $\lambda = \bar{q}L/T_0$.

The integration of the finite difference method within the SPH framework was introduced in Bai's research (Bai et al., 2018). Here, differential equations were employed during post-processing of SPH simulation to compute the temperature gradient. For complex fluid-structure coupling boundaries, various enhanced computational techniques have also been developed, further improving simulation accuracy.

5.3.3 Verification of Typical Cases

5.3.3.1 Case 1: Homogeneous Medium

For this simulation, the infinite plate is treated as a homogeneous medium with the following properties: density $\rho = 1000\,\mathrm{kg/m^3}$, specific heat capacity $c = 1000\,\mathrm{J/(kg\cdot °C)}$, and default thermal conductivity $\lambda_0 = 1\,\mathrm{W/(m\cdot °C)}$. The initial temperature of the particles is set to 0. In this calculation, the boundary conditions are applied as shown in Figure 5.17. The thickness of the square medium is $L = 0.012\,\mathrm{m}$. Other computational parameters include a particle spacing $\Delta x = 0.0002\,\mathrm{m}$, particle number $N = 3600$, a temperature difference $T_0 = 1\,°\mathrm{C}$, and a time step $\Delta t = 0.0001\,\mathrm{s}$.

From Figure 5.18a, the temperature distribution in the porous medium gradually converges toward a linear profile. The dimensionless time factor is defined as $t_\mathrm{d} = \kappa t/L^2$, with κ being the thermal diffusion coefficient of water. Figure 5.18b demonstrates that the calculated thermal conductivity rapidly approaches the baseline value with increasing t_d.

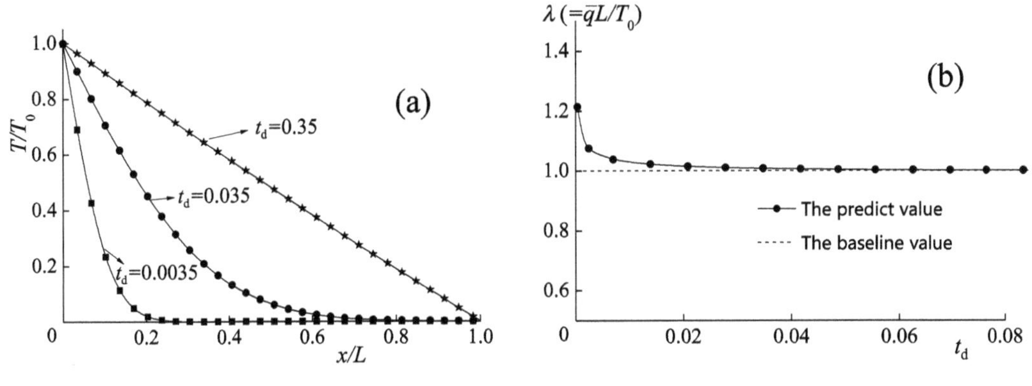

FIGURE 5.18 Calculation results. (a) Temperature distribution. (b) $\lambda - t_d$ relation.

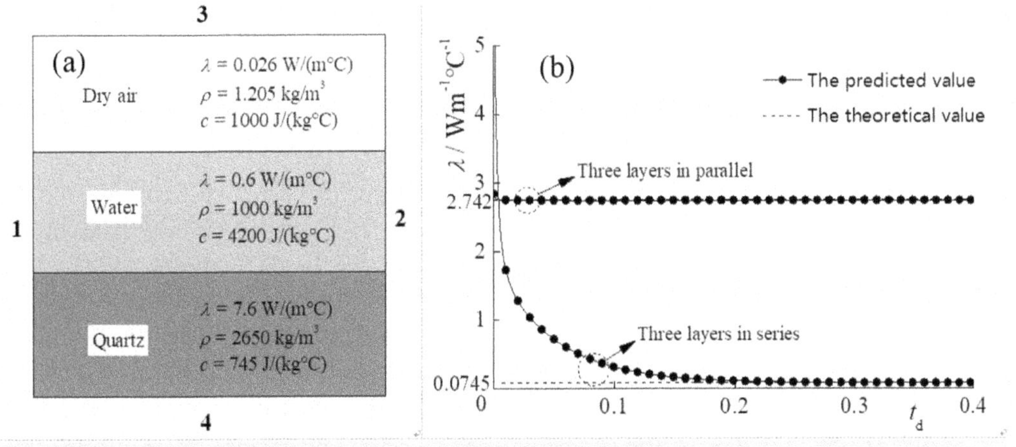

FIGURE 5.19 Three-phase layered medium. (a) Sketch map. (b) Variation of λ with t_d.

TABLE 5.1 Thermal Parameters of the Porous Medium

Materials	λ W·(m⁻¹·°C⁻¹)	ρ kg·m⁻³	c J·(kg⁻¹·°C⁻¹)	κ m²·s⁻¹
Solid: quartz (20 °C)	7.6	2650	745	3.85×10^{-6}
Dry air (20 °C, atmospheric pressure)	0.026	1.20	1000	2.17×10^{-5}
Water (20 °C)	0.6	1000	4200	1.43×10^{-7}

5.3.3.2 Case 2: Three-Phase Layered Medium

As illustrated in Figure 5.19a, the solid, liquid, and gas phases are assumed to be arranged in distinct layers, with each phase occupying an equal volume within the medium. The materials properties are provided in Table 5.1. When boundaries 1 and 2 are defined as periodic boundaries and boundaries 3 and 4 as temperature boundaries, the three layers are effectively connected in series. Conversely, if boundaries 3 and 4 are periodic while boundaries 1 and 2 are temperature boundaries, the layers are connected in parallel. The simulation parameters remain consistent with those used in Case 1. As indicated by Eq. (5.21), the theoretical effective thermal conductivity (λ) for the three-phase series configuration is 0.0745 W/(m·°C), while that for the parallel arrangement is 2.742 W/(m·°C).

Figure 5.19b demonstrates that, under both connection modes, the predicted effective thermal conductivity values (λ) converge accurately to the corresponding theoretical values within the selected simulation period. The results confirm that the proposed simulation approach is reliable for predicting the effective thermal conductivities of multiphase media.

5.3.4 Thermal Conductivity with Saturation Degree

In the SAP1 simulation, the upper boundary (plate $z = L$) is assigned a temperature of T_0, while the lower boundary (plate $z = 0$) is maintained at 0. The remaining boundaries are configured as periodic, effectively modeling the granular medium as an infinite porous layer of thickness L. Material properties are assigned based on the values in Table 5.1, and the time step is set to $\Delta t = 0.0001$ s. Using this setup, the steady-state temperature distribution of the specimen is successfully predicted, as shown in Figure 5.20a.

From Figure 5.20b, the calculated effective thermal conductivity progressively stabilizes over time. To ensure the precision of the predicted λ value to three significant figures, the simulation continues until the fourth significant digit remains unchanged over a time interval of $\Delta t_d = 0.01$. From Figure 5.20b, the final result for the effective thermal conductivity for the medium reaches $\lambda = 2.75$ W/(m·°C), in agreement with the Wiener limits (the upper bound being 4.32 W/(m·°C) and the lower bound being 0.06 W/(m·°C).

A control specimen designated SAP2 (Figure 5.21), featuring a granule radius of $a = 0.8$ mm, is generated with the same porosity as SAP1. By applying the humidifying algorithm to both SAP1 and SAP2, two series of specimens with progressively increasing saturation are obtained. In the simulation, the solid phase for each soil specimen

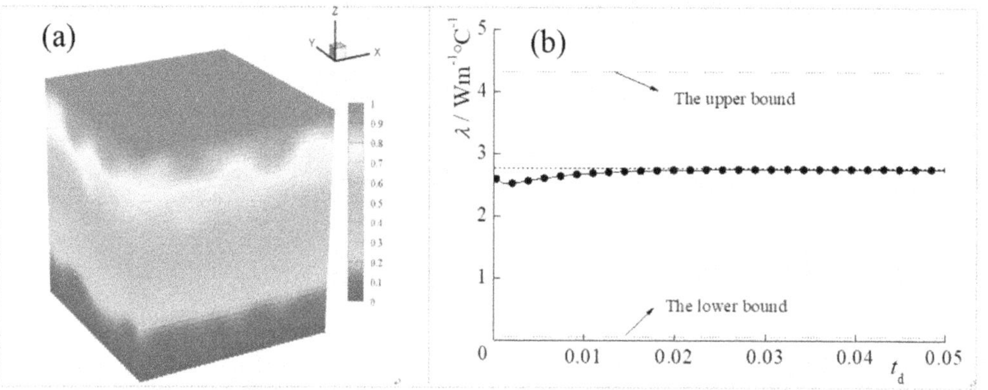

FIGURE 5.20 3D stacked medium. (a) Temperature distribution. (b) Predicted thermal conductivity.

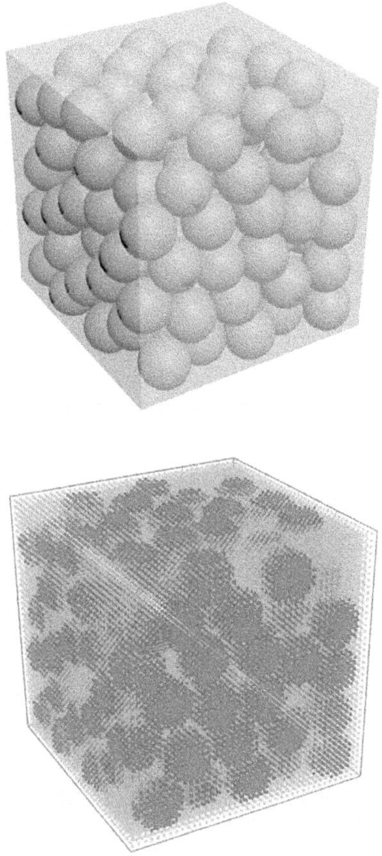

FIGURE 5.21 Particle distribution. (a) Control specimen. (b) Discretization of SAP2.

group is successively assigned the properties of quartz and then feldspar, thereby yielding four distinct sets of simulation results. Thus, four sets of simulation results are obtained. For feldspar, the values of λ, ρ, and c are 2.3 W/(m·°C), 2650 kg/m^3, and 710 J/(kg·°C), respectively.

Cote and Konrad (2005) established an empirical equation to describe the thermal conductivity:

$$\frac{1}{\lambda_r} = \frac{1}{\alpha} \cdot \frac{1}{S_r} + \left(1 - \frac{1}{\alpha}\right) \tag{5.23}$$

where λ_r is the normalized thermal conductivity, S_r is the saturation degree of porous medium, and α is an influence coefficient.

The value of λ_r varies between 0 and 1, which is related to the types of soils (Bai et al., 2022b). From the previous calculation results, the evolution curve for λ_r versus S_r is plotted in Figure 5.22. The results indicate that the slope of the $\lambda_r - S_r$ curve attains its maximum at zero saturation and subsequently diminishes gradually. This trend suggests that the

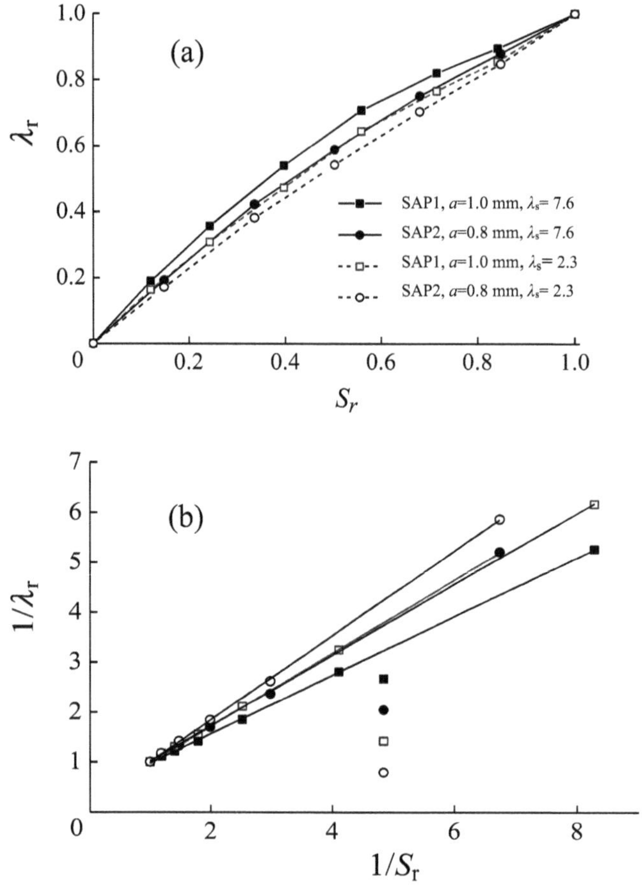

FIGURE 5.22 The relationship between λ_r and S_r. (a) Evolution curve. (b) Linear relation.

establishment of a liquid bridge during the initial humidifying stage enhances the thermal conductivity of the porous media.

From Figure 5.22b, a strong linear correlation exists between $1/\lambda_r$ and $1/S_r$, which is consistent with the empirical relationship given in Eq. (5.23). Furthermore, the comparison of the solid and hollow dots in Figure 5.22 reveals that the slope of the linear fit for the $1/\lambda_r$ - $1/S_r$ plot varies with both material type and granule size. Therefore, the influence coefficient α in Eq. (5.23) is dependent on the thermal conductivity of the solid phase and the granular characteristics of the medium.

CHAPTER 6

Theoretical Methods of Suspended Substance Transport

6.1 MAIN CONTENT OVERVIEW

For a long-time injection case, the mobilization of suspended substances (e.g., suspended particles [SPs]; heavy metal ions [HMs]) becomes increasingly significant. To describe this behavior, a governing equation that incorporates the detachment of the absorbed particles is formulated, capturing one-dimensional flow and dispersion through the application of a source-function approach. By employing Laplace and Fourier transformations and subsequently performing inverse transforms, an elementary solution is obtained for an instantaneous point source introduced at the surface of a porous medium. Building on this foundation, integral techniques are applied to derive analytical expressions for contaminant concentration within the porous domain when the system is influenced by a localized surface source in which spatial configuration and temporal strength change arbitrarily over the free boundary.

The transport parameters, including dispersivity, attachment (deposition or adsorption) coefficient, and detachment (or release) coefficient, are estimated on the basis of previously obtained experimental data. Using these calibrated parameters, the developed model is further employed to predict the transport behavior of SPs under different environmental settings and hydrodynamic regimes.

Moreover, the cotransport of SPs and HMs is examined under a range of conditions, such as variations in temperature, seepage velocity, SPs size, and HMs injection concentration. By distinguishing the variance in particle size and dielectric property, a theoretical cotransport model for SPs and HMs based on the convection–diffusion–attachment hydrodynamics of suspended substances is proposed and subsequently verified against experimental data.

DOI: 10.1201/9781003783749-6

6.2 SOURCE-FUNCTION METHOD OF A ONE-DIMENSIONAL PROBLEM

6.2.1 Governing Equation and the Solution in Transform Space

The transport of suspended substances within a uniform one-dimensional steady flow directed along the z-axis is formulated by considering the detachment associated with hydrodynamic attachment. The corresponding partial differential equation describes the combined effects of advection, dispersion, and the detachment of suspended substances previously deposited within the porous medium (Bai et al., 2016):

$$\left.\begin{aligned} &\frac{\partial C(z,t)}{\partial t}=D\frac{\partial^2 C(z,t)}{\partial z^2}-u\frac{\partial C(z,t)}{\partial z}-\frac{\rho_s}{n}\frac{\partial \sigma(z,t)}{\partial t} \\ &\frac{\rho_s}{n}\frac{\partial \sigma(z,t)}{\partial t}=k_d\cdot C(z,t)-k_r\cdot\frac{\rho_s}{n}\cdot\sigma(z,t) \end{aligned}\right\} \tag{6.1}$$

where n represents the porosity of the porous medium, t indicates the elapsed time, D represents the hydrodynamic dispersion coefficient, z denotes the spatial coordinate, u refers to the average interstitial velocity, σ corresponds to the concentration of suspended substances retained on the solid matrix, ρ_s signifies the bulk density of the solid matrix, and k_d and k_r describe the attachment coefficient and the detachment coefficient, respectively.

The three terms on the right side of Eq. (6.1) describe the contribution of hydrodynamic dispersion, the convective transport arising from hydraulic permeability, and the attachment behavior of suspended substances that includes their subsequent detachment. When the porous matrix is assumed to contain no absorbed suspended substances at the start, and suspended substances with concentrations that vary over time are supplied through the top boundary of the column, the system is defined by the following initial and boundary conditions: $C(z,0)=0$, $\sigma(z,0)=0$, $C(0,t)=g(t)$, and $C(+\infty,t)=0$.

The second expression in Eq. (6.1) $(\sigma(z,t))$ can be evaluated through the application of the Laplace transform method:

$$\sigma(z,t)=k_d\cdot\frac{n}{\rho_s}\cdot\int_0^t C(z,\tau)\cdot\exp\left[-k_r(t-\tau)\right]\mathrm{d}\tau \tag{6.2}$$

Using the Leibniz integral rule, the differentiation of Eq. (6.2) with respect to t is written as follows:

$$\frac{\partial \sigma(z,t)}{\partial t}=-k_r\cdot k_d\cdot\frac{n}{\rho_s}\cdot\int_0^t C(z,\tau)\cdot\exp\left[-k_r(t-\tau)\right]\mathrm{d}\tau+k_d\cdot\frac{n}{\rho_s}\cdot C(z,t) \tag{6.3}$$

By inserting Eq. (6.3) into the first expression of Eq. (6.1), one obtains:

$$\begin{aligned} \frac{\partial C(z,t)}{\partial t}&=D\frac{\partial^2 C(z,t)}{\partial z^2} \\ &-u\frac{\partial C(z,t)}{\partial z}+k_r\cdot k_d\cdot\int_0^t C(z,\tau)\cdot\exp[-k_r(t-\tau)]\mathrm{d}\tau-k_d\cdot C(z,t) \end{aligned} \tag{6.4}$$

The Laplace transforms of the function with respect to t and z are given by $\bar{C}=\int_0^{\infty}C(z,t)e^{-st}\,\mathrm{d}t$ and $\tilde{\bar{C}}=\int_0^{\infty}\bar{C}(z,s)e^{-rz}\,\mathrm{d}z$, respectively, where r and s denote the Laplace transform variables. Noting that $C(z,\,0)=0$ and $\bar{C}(0,s)=\bar{g}(s)$, applying the Laplace transform first in t and subsequently in z to Eq. (6.4) yields

$$\tilde{\bar{C}}(r,s)=\frac{\bar{g}\cdot r-\dfrac{u}{D}\cdot\bar{g}+\dfrac{\mathrm{d}\bar{C}(0,s)}{\mathrm{d}z}}{(r+A)^2-B^2}\tag{6.5}$$

where $A=-u/(2D)$, $B=[A^2+(s+\dfrac{k_{\mathrm{d}}\cdot s}{s+k_{\mathrm{r}}})/D]^{1/2}$.

6.2.2 Inversion of the Mathematical Transform

The inversion of the Laplace transform in Eq. (6.5) in terms of r is (Abramowitz and Stegun, 1972):

$$\begin{aligned}\bar{C}(z,s)=&\frac{1}{B}\left[-\frac{u}{D}\cdot\bar{g}+\frac{\mathrm{d}\bar{C}(0,s)}{\mathrm{d}z}\right]\cdot\exp(-Az)\cdot\sinh(Bz)\\&+\bar{g}\cdot\frac{\exp(-Az)}{B}\cdot[B\cdot\cosh(Bz)-A\sinh(Bz)]\end{aligned}\tag{6.6}$$

By multiplying both sides of Eq. (6.6) by $2B\cdot\exp[(A-B)z]$ and applying $C(+\infty,t)=0$, the following expression is obtained:

$$\frac{\mathrm{d}\bar{C}(0,s)}{\mathrm{d}z}=\frac{u}{D}\cdot\bar{g}+(A-B)\cdot\bar{g}\tag{6.7}$$

Introducing Eq. (6.7) into Eq. (6.6) leads to the following result:

$$\bar{C}(z,s)=\bar{g}\cdot\exp(-Az)\cdot\exp(-Bz)\tag{6.8}$$

In Eq. (6.8), $\exp(-Bz)$ can be written as

$$h(z,s)=\exp(-Bz)=\exp\left[-\frac{z}{\sqrt{D(s+k_{\mathrm{r}})}}(s^2+s(k_{\mathrm{r}}+E)+k_{\mathrm{r}}E-k_{\mathrm{r}}\cdot k_{\mathrm{d}})^{\frac{1}{2}}\right]\tag{6.9}$$

where $E=DA^2+k_{\mathrm{d}}$.

Based on Abramowitz and Stegun (1972) as well as Sim and Chrysikopoulos (1998), one obtains the following relationship:

$$L^{-1}[\frac{1}{s+k_{\mathrm{r}}}\bar{f}_0(s+k_{\mathrm{r}}-\frac{\alpha}{s+k_{\mathrm{r}}})]=\exp^{-k_{\mathrm{r}}t}\int_0^t I_0[2(\alpha\eta(t-\eta))^{1/2}]f_0(\eta)\,\mathrm{d}\eta\tag{6.10}$$

Here L^{-1} denotes the inverse Laplace operator, α represents an arbitrary constant, I_0 represents the modified Bessel function of the first kind of order zero, and $\overline{f}_0(s)$ corresponds to the Laplace transform of an arbitrary function $f_0(t)$, which is defined as follows:

$$f_0(t) = \frac{\alpha_1}{(4\pi t^3)^{1/2}} \exp(-\frac{\alpha_1^2}{4t} - \alpha_2 t) \tag{6.11}$$

where α_1 and α_2 denote the arbitrary constants.

The Laplace transformation of Eq. (6.11) in the time variable t can be written as follows:

$$\overline{f}_0(s) = \exp\left[-\alpha_1 (s+\alpha_2)^{1/2}\right] \tag{6.12}$$

It follows that

$$\overline{f}_0(s + k_r - \frac{\alpha}{s+k_r}) = \exp\{-\frac{\alpha_1}{(s+k_r)^{1/2}}[s^2 + s(2k_r + \alpha_2) + k_r(k_r + \alpha_2) - \alpha]^{1/2}\} \tag{6.13}$$

Suppose $\alpha = k_r k_d$, $\alpha_1 = z/\sqrt{D}$, $\alpha_2 = E - k_r$. A comparison between Eqs. (6.9) and (6.13) leads to

$$h(z,s) = \overline{f}_0(s + k_r - \frac{\alpha}{s+k_r}) \tag{6.14}$$

Accordingly, Eq. (6.10) can be rewritten as

$$L^{-1}[\frac{h(z,s)}{s+k_r}] = \exp^{-k_r t} \int_0^t I_0[2(\alpha\eta(t-\eta))^{1/2}] f_0(\eta)\, d\eta \tag{6.15}$$

Let $R(z,t) = L^{-1}[h(z,s)/(s+k_r)]$. By applying the derivative property of the Laplace transform and using $R(z,0) = 0$, it follows that $\partial R(z,t)/\partial t = L^{-1}[s \cdot h(z,s)/(s+k_r)] = L^{-1}[s\overline{R}(z,s)]$.

From Eq. (6.9), one obtains

$$L^{-1}[\exp(-Bz)] = L^{-1}[\frac{k_r \cdot h(z,s)}{s+k_r} + \frac{s \cdot h(z,s)}{s+k_r}] = k_r \cdot R(z,t) + \frac{\partial R(z,t)}{\partial t} \tag{6.16}$$

From Eqs. (6.11) and (6.16), it can be derived that

$$R(z,t) = \exp^{-k_r t} \int_0^t I_0[2(\alpha\eta(t-\eta))^{1/2}] \cdot \frac{\alpha_1}{2\sqrt{\pi\eta^3}} \cdot \exp[-\frac{\alpha_1^2}{4\eta} - \alpha_2\eta]\, d\eta \tag{6.17}$$

$$\frac{\partial R(z,t)}{\partial t} = \exp^{-k_r t} \int_0^t \{(\frac{\alpha\eta}{t-\eta})^{1/2} \cdot I_1[2(\alpha\eta(t-\eta))^{1/2}] - k_r I_0[2(\alpha\eta(t-\eta))^{1/2}]\}$$
$$\cdot \frac{\alpha_1}{2\sqrt{\pi\eta^3}} \exp[-\frac{\alpha_1^2}{4\eta} - \alpha_2\eta]\, d\eta + \exp^{-k_r t} \frac{\alpha_1}{2\sqrt{\pi t^3}} \exp[-\frac{\alpha_1^2}{4t} - \alpha_2 t] \tag{6.18}$$

where I_1 denotes the modified Bessel function of the first kind of order one.

By applying the convolution theorem, the concentration of the suspended substance can be derived from Eq. (6.8):

$$C(z,t)=\exp(\frac{uz}{2D})\cdot\int_0^t g(\tau)\cdot[k_r\cdot W(z,t-\tau)+\frac{\partial W(z,t-\tau)}{\partial t}]\mathrm{d}\tau \tag{6.19}$$

where τ refers to a dummy integration variable.

$$W(z,t)=\exp^{-k_r t}\int_0^t I_0[2(\alpha\eta(t-\eta))^{1/2}]\cdot\frac{z}{2\eta\sqrt{\pi D\eta}}\cdot\exp[-\frac{z^2}{4D\eta}-\frac{u^2\eta}{4D}+(k_r-k_d)\eta]\mathrm{d}\eta \tag{6.20}$$

$$\frac{\partial W(z,t)}{\partial t}=\exp^{-k_r t}\int_0^t\left\{(\frac{\alpha\eta}{t-\eta})^{1/2}I_1[2(\alpha\eta(t-\eta))^{1/2}]-k_r I_0[2(\alpha\eta(t-\eta))^{1/2}]\right\}$$
$$\cdot\frac{z}{2\eta\sqrt{\pi D\eta}}\cdot\exp[-\frac{z^2}{4D\eta}-\frac{u^2\eta}{4D}+(k_r-k_d)\eta]\mathrm{d}\eta+\exp^{-k_r t}\cdot\frac{z}{2t\sqrt{\pi Dt}}\cdot\exp[-\frac{z^2}{4Dt}-\frac{u^2 t}{4D}+(k_r-k_d)t] \tag{6.21}$$

6.2.3 Solution of a Continuous Injection over Time

For an instantaneous plane source applied at the top surface of the column, where the suspended substance concentration is $g(t)$ $[\mathrm{ML}^{-3}]$, the following expression is obtained:

$$g(t)=I\cdot\delta(t-t') \tag{6.22}$$

where $I=m/Q=m/(vA)$ represents the strength of the plane source $[\mathrm{MTL}^{-3}]$, where m is the injected suspended substance mass [M], Q is the volumetric flow rate $[L^3T^{-1}]$, $v=n\cdot u$ denotes the seepage velocity, and A is the cross-sectional area of the column $[L^2]$. $\delta(\cdot)$ is the Dirac delta function, and t' refers to the injection time [T].

It is noted that $\int_\alpha^\beta f(t)\delta(t-t_0)\mathrm{d}t=f(t_0)\,(\alpha\le t_0\le\beta)$ substituting Eq. (6.22) into Eq. (6.19) yields

$$C(z,t-t')=\frac{I}{2\sqrt{\pi D}}\cdot A_1(z,t-t')$$
$$\left\{\int_0^{t-t'}A_2(t-t')\cdot A_3(z,\eta)\cdot A_4(z,\eta)\mathrm{d}\eta+A_3(z,t-t')\cdot A_4(z,t-t')\right\} \tag{6.23}$$

where $A_1(z,t)=\exp(\frac{uz}{2D}-k_r t)$, $A_2(t)=(\frac{\alpha\eta}{t-\eta})^{1/2}\cdot I_1[2(\alpha\eta(t-\eta))^{1/2}]$, $A_3(z,t)=\frac{z}{t^{3/2}}$, and $A_4(z,t)=\exp[-\frac{z^2}{4Dt}-\frac{u^2t}{4D}+(k_r-k_d)t]$.

When $I = 1$, the fundamental solution corresponding to an instantaneous planar source can be derived from Eq. (6.23) as follows:

$$C_P(z,t-t') = \frac{1}{2\sqrt{\pi D}} \cdot A_1(z,t-t')$$
$$\left\{ \int_0^{t-t'} A_2(t-t') \cdot A_3(z,\eta) \cdot A_4(z,\eta) \mathrm{d}\eta + A_3(z,t-t') \cdot A_4(z,t-t') \right\} \tag{6.24}$$

The traditional analytical solution derived from the classical convection–dispersion framework, which neglects the influence of the detachment effect on adsorption-kinetics, can be recovered by assigning $k_r = 0$ and $t' = 0$ in Eq. (6.23). Such formulations are commonly employed to simulate breakthrough curves (BTCs) obtained from column experiments under pulse-injection conditions, where the duration of the applied pulse is sufficiently brief to be treated as instantaneous. Nevertheless, these theoretical descriptions exhibit growing discrepancies from experimental observations as the injection period becomes longer. For scenarios involving sustained injection, it is therefore essential to develop a solution that incorporates the detachment behavior of adsorbed suspended substances, accounting for the prolonged detachment process associated with suspended substances.

For a continuously applied concentration $g(t)$ over time t, the transport behavior may be conceptualized as the superposition of many infinitesimal instantaneous fictitious planar sources with strengths $I = g(t)\mathrm{d}t$ (Bai, 2006). By integrating Eq. (6.24) with respect to time, the resulting suspended substance concentration can be expressed as follows:

$$C(z,t) = \int_0^t C_P(z,t-t') \cdot g(t') \cdot \mathrm{d}t' \tag{6.25}$$

When the concentration $g(t)$ is maintained at a constant value g throughout the injection period, the suspended substance concentration within the porous medium can be directly determined from Eq. (6.25).

6.2.4 Proposed Model Verification

As discussed in Section 6.2.3, the pulse-injection column experiment and its corresponding analytical solution provide an effective means for determining the transport parameters of SPs. Figure 6.1a illustrates the calculated BTCs (i.e., the relationship between relative concentration C_R and pore volume V_P; see Chapter 2) for a range of injection durations ($t_{inj} = 0.2,\ 1,\ 2,\ 5,\ 10,\ 30$ s) and demonstrates that the peak of each BTCs shifts to later times as the injection interval increases. Moreover, the predicted peak concentrations diminish progressively with longer injection periods when compared with those associated with an instantaneous input (i.e., $t_{inj} \to 0$; Figure 6.1b). The relative discrepancies between the two cases remain minimal only when the injection duration is extremely short; for instance, the error remains below 1% for $t_{inj} < 1$ s at $D_{50} = 1.02$ μm.

Figure 6.2 presents both the measured and simulated BTCs (i.e., effluent concentration C_{out} versus V_P) for short-time injections (i.e., $t_{inj} = 30$, 60, and 120 s) of SPs with a median diameter of $D_{50} = 10$ μm under two seepage conditions, $v = 0.087$ and 0.260 cm/s. The injection concentration in these tests is $C_{inj} = 3.75$ mg/ml, and the injected volume

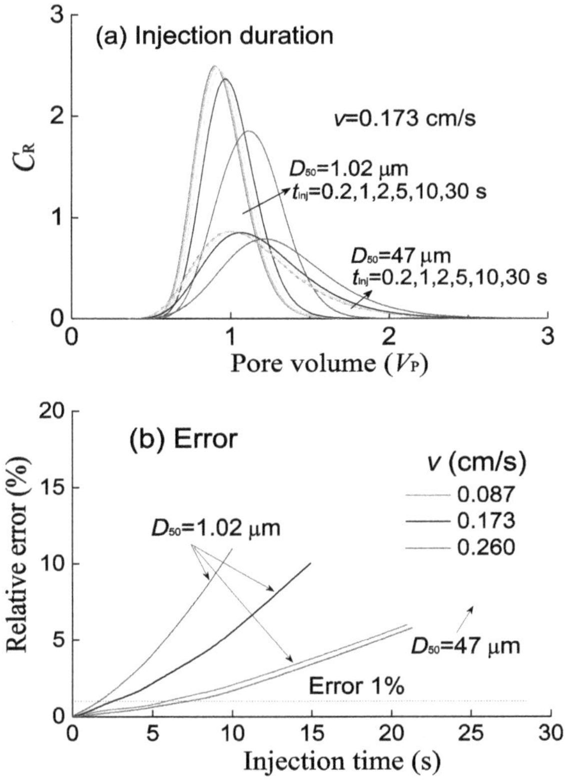

FIGURE 6.1 Effects of injection time. (a) Injection duration. (b) Error.

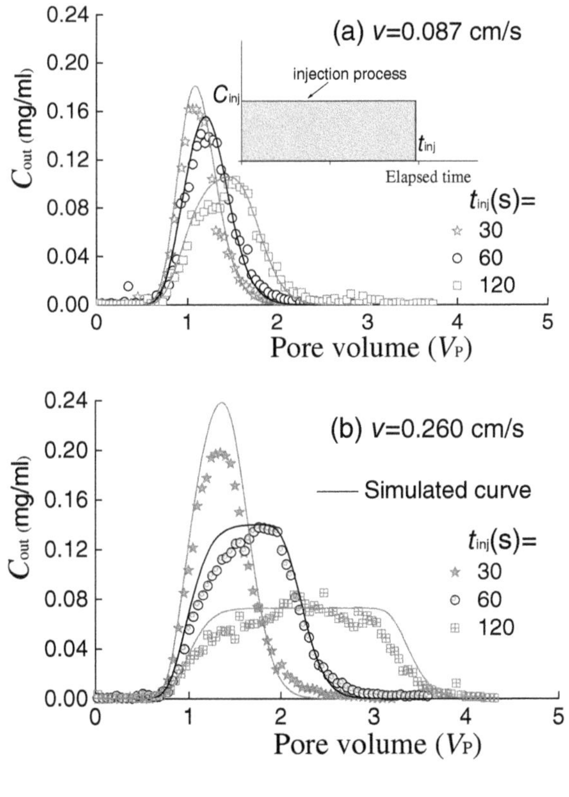

FIGURE 6.2 BTCs for short-time injection. (a) v = 0.087 cm/s. (b) v = 0.260 cm/s.

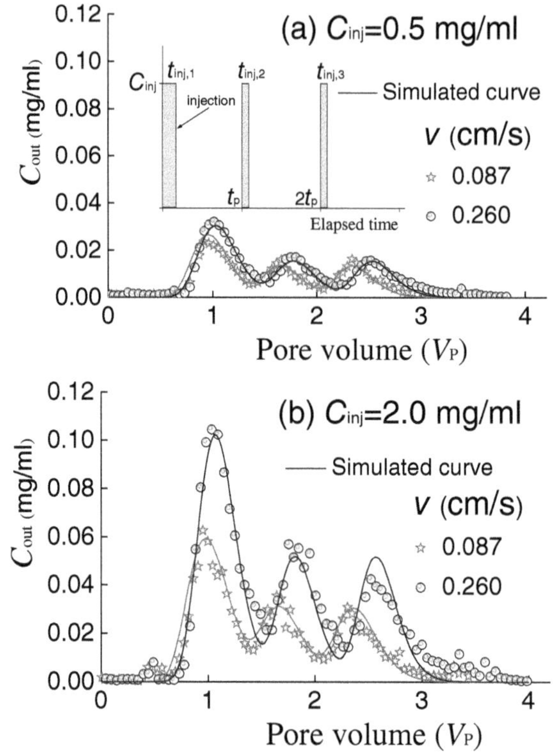

FIGURE 6.3 BTCs for three-pulse injections. (a) $C_{\text{inj}} = 0.5$ mg/ml. (b) $C_{\text{inj}} = 2.0$ mg/ml.

TABLE 6.1 Fitted Parameters Associated with the Injection Cases

Injection pattern	Injection concentration C_{inj} (mg/ml)	$v = 0.087$ cm/s			$v = 0.260$ cm/s		
		u_r	α_d(cm)	k_d(s^{-1})	u_r	α_d(cm)	k_d(s^{-1})
Three pulse	0.5	1.0	0.55	0.004	0.95	0.5	0.008
Three pulse	1	1.0	0.7	0.006	0.95	0.5	0.011
Three pulse	2	0.95	0.6	0.007	0.92	0.4	0.012
Short-time	3.75	0.96	0.6	0.008	0.96	0.5	0.012

Note: u_r, α_d, and k_d denote relative interstitial particle velocity to water velocity, dispersivity, and attachment coefficient, respectively.

is $V_{\text{inj}} = 60$ ml. As previously noted, determining transport parameters is challenging because they depend on the factors such as particle size, concentration, and seepage velocity. Figure 6.2 shows that the proposed model provides a satisfactory match to the experimental results.

Figure 6.3 presents both the experimental and predicted BTCs corresponding to a sequence of three repeated pulse-injections ($t_{\text{inj},1} = 2$ s, $t_{\text{inj},2} = 1$ s, and $t_{\text{inj},3} = 1$s) for SPs with a median size of $D_{50} = 10$ μm, evaluated at three different particle concentrations. The time intervals between successive pulses are $t_p = 104$ s for $v = 0.087$ cm/s and $t_p = 38$ s for 0.260 cm/s. The transport parameters used in the simulations are listed in Table 6.1.

The comparison demonstrates that the proposed model captures the observed behavior effectively and appears suitable for representing transport under more complex, variable-concentration injection scenarios.

6.3 SOURCE-FUNCTION METHOD IN A SEMI-INFINITE POROUS MEDIUM

6.3.1 Governing Equation and the Solution

6.3.1.1 Governing Equations

It is assumed that contaminants are introduced through a localized region on the surface of a saturated, homogeneous half-space porous medium, which extends infinitely in the *x*-axis and *y*-axis directions and is semi-infinite along the *z*-axis. The transport of solutes under one-dimensional steady flow in the *z* direction, while incorporating the effects of three-dimensional hydrodynamic dispersion, is described by the following partial differential equation (Bai et al., 2015):

$$\begin{aligned}\frac{\partial C(x,y,z,t)}{\partial t} &= D_x\frac{\partial^2 C(x,y,z,t)}{\partial x^2}+D_y\frac{\partial^2 C(x,y,z,t)}{\partial y^2}+D_z\frac{\partial^2 C(x,y,z,t)}{\partial z^2}\\ &-u\frac{\partial C(x,y,z,t)}{\partial z}-\frac{\rho_s}{n}\frac{\partial \sigma(x,y,z,t)}{\partial t}\end{aligned} \tag{6.26}$$

where x and y denote the spatial coordinates orthogonal to the flow direction, z represents the coordinate aligned with the water flow, n denotes the porosity of the medium, C refers to the solute concentration in the liquid phase, the coefficients D_x, D_y and D_z correspond to the hydrodynamic dispersion in the x, y, and z directions, respectively, u is the average interstitial velocity, and t denotes the time, σ signifies the amount of solute retained on the solid matrix, and ρ_s represents the bulk density of the solid phase.

The first three terms on the right-hand side of Eq. (6.26) describe hydrodynamic dispersion occurring in the *x*, *y*, and *z* directions; the fourth term corresponds to convective transport associated with hydraulic permeability; and the fifth term characterizes the dynamic attachment of contaminants, including the detachment process, and can be expressed as follows:

$$\frac{\rho_s}{n}\cdot\frac{\partial \sigma(x,y,z,t)}{\partial t}=k_d\cdot C(x,y,z,t)-k_r\cdot\frac{\rho_s}{n}\cdot\sigma(x,y,z,t) \tag{6.27}$$

where k_d and k_r denote the attachment and detachment coefficients associated with the solute in the liquid phase and the adsorbed phase, respectively.

6.3.1.2 Solutions in the Transform Space

Assuming that the porous medium initially contains no adsorbed solutes, the solute input is applied at the surface of the semi-infinite domain, with concentrations that vary in both time and spatial coordinates. Under these conditions, the governing equation is supplemented with the following initial and boundary specifications: $C(x,y,z,0)=0$, $\sigma(x,y,z,0)=0$, $C(x,y,0,t)=g(x,y,t)$, $C(x,y,+\infty,t)=0$, $C(\pm\infty,y,z,t)=0$, and $C(x,\pm\infty,z,t)=0$.

The $\sigma(x,y,z,t)$ in Eq. (6.27) can be readily evaluated by applying the Laplace transform method, yielding

$$\sigma(x,y,z,t)=k_{\mathrm{d}}\cdot\frac{n}{\rho_{\mathrm{s}}}\cdot\int_0^t C(x,y,z,\tau)\cdot\exp\left[-k_{\mathrm{r}}(t-\tau)\right]d\tau \tag{6.28}$$

By applying the Leibniz integral rule, the time derivative of Eq. (6.28) can be expressed as follows:

$$\frac{\partial\sigma(x,y,z,t)}{\partial t}=-k_{\mathrm{r}}\cdot k_{\mathrm{d}}\cdot\frac{n}{\rho_{\mathrm{s}}}\cdot\int_0^t C(x,y,z,\tau)\cdot\exp\left[-k_{\mathrm{r}}(t-\tau)\right]d\tau+k_{\mathrm{d}}\cdot\frac{n}{\rho_{\mathrm{s}}}\cdot C(x,y,z,t) \tag{6.29}$$

By inserting Eq. (6.29) into Eq. (6.26), the following expression is obtained:

$$\begin{aligned}\frac{\partial C(x,y,z,t)}{\partial t}&=D_x\frac{\partial^2 C(x,y,z,t)}{\partial x^2}+D_{\mathrm{y}}\frac{\partial^2 C(x,y,z,t)}{\partial y^2}+D_z\frac{\partial^2 C(x,y,z,t)}{\partial z^2}\\&\quad-u\frac{\partial C(x,y,z,t)}{\partial z}+k_{\mathrm{r}}\cdot k_{\mathrm{d}}\cdot\int_0^t C(x,y,z,\tau)\cdot\exp[-k_{\mathrm{r}}(t-\tau)]d\tau-k_{\mathrm{d}}\cdot C(x,y,z,t)\end{aligned} \tag{6.30}$$

The Laplace transforms of the function with respect to time t and the spatial coordinate z are denoted by $\bar{C}=\int_0^\infty C(x,y,z,t)e^{-st}\,dt$ and $\bar{\bar{C}}=\int_0^\infty \bar{C}(x,y,z,s)e^{-rz}\,dz$, respectively, where s and r serve as the associated Laplace variables. Since $C(x,y,z,\ 0)\ =\ 0$, applying the Laplace transform in the time domain to Eq. (6.30) leads to the following expression:

$$s\cdot\bar{C}=D_x\frac{\partial^2\bar{C}}{\partial x^2}+D_y\frac{\partial^2\bar{C}}{\partial y^2}+D_z\frac{\partial^2\bar{C}}{\partial z^2}-u\frac{\partial\bar{C}}{\partial z}+\frac{k_{\mathrm{r}}\cdot k_{\mathrm{d}}\cdot\bar{C}}{s+k_{\mathrm{r}}}-k_{\mathrm{d}}\cdot\bar{C} \tag{6.31}$$

In a similar manner, noting that $\bar{C}(x,y,0,s)=\bar{g}(x,y,s)$, applying the Laplace transform in the z domain to Eq. (6.31) results in

$$\begin{aligned}(s+k_{\mathrm{d}})\cdot\bar{\bar{C}}&=D_x\frac{\partial^2\bar{\bar{C}}}{\partial x^2}+D_y\frac{\partial^2\bar{\bar{C}}}{\partial y^2}+D_z\left(r^2\bar{\bar{C}}-r\bar{g}-\frac{\partial\bar{C}(x,y,0,s)}{\partial z}\right)\\&\quad-u(r\bar{\bar{C}}-\bar{g})+\frac{k_{\mathrm{r}}\cdot k_{\mathrm{d}}\cdot\bar{\bar{C}}}{s+k_{\mathrm{r}}}\end{aligned} \tag{6.32}$$

Applying the Fourier transform in the x and y directions to Eq. (6.32) leads to

$$\begin{aligned}(s+k_{\mathrm{d}})\cdot\bar{\bar{C}}^{xy}&=D_z\left(r^2\bar{\bar{C}}^{xy}-r\bar{g}^{xy}-\frac{\partial\bar{C}^{xy}(x,y,0,s)}{\partial z}\right)-D_x\omega^2\bar{\bar{C}}^{xy}-D_y\phi^2\bar{\bar{C}}^{xy}\\&\quad-u(r\bar{\bar{C}}^{xy}-\bar{g}^{xy})+\frac{k_{\mathrm{r}}\cdot k_{\mathrm{d}}\cdot\bar{\bar{C}}^{xy}}{s+k_{\mathrm{r}}}\end{aligned} \tag{6.33}$$

where ω and ϕ represent the Fourier transform variables of x and y, respectively.

Equation (6.33) can be reformulated as follows:

$$\tilde{\bar{C}}^{xy} = \frac{\bar{g}^{xy} \cdot r - \dfrac{u}{D_z} \cdot \bar{g}^{xy} + \dfrac{\mathrm{d}\bar{C}^{xy}(w,\phi,0,s)}{\mathrm{d}z}}{(r+A)^2 - B^2} \tag{6.34}$$

where $A = -u/(2D_z)$ and $B = [A^2 + (D_x w^2 + D_y \phi^2 + s + \dfrac{k_\mathrm{d} \cdot s}{s+k_\mathrm{r}})/D_z]^{1/2}$.

6.3.1.3 Inversion of the Mathematical Transform

The inversion of the Laplace transform in Eq. (6.34) with respect to the variable r can be expressed as follows:

$$\begin{aligned}\bar{C}^{xy} = &\frac{1}{B}\left[-\frac{u}{D_z} \cdot \bar{g}^{xy} + \frac{\mathrm{d}\bar{C}^{xy}(w,\theta,0,s)}{\mathrm{d}z}\right] \cdot \exp(-Az) \cdot \sinh(Bz) \\ &+ \bar{g}^{xy} \cdot \frac{\exp(-Az)}{B} \cdot [B \cdot \cosh(Bz) - A \sinh(Bz)]\end{aligned} \tag{6.35}$$

Multiplying both sides of Eq. (6.35) by $2B \cdot \exp[(A-B)z]$ simultaneously gives the following expression:

$$\begin{aligned}2B \cdot \exp[(A-B)z] \cdot \bar{C}^{xy} = &\left[-\frac{u}{D_z} \cdot \bar{g}^{xy} + \frac{\mathrm{d}\bar{C}^{xy}(w,\theta,0,s)}{\mathrm{d}z}\right] \cdot [1 - \exp(-2Bz)] \\ &+ \bar{g}^{xy} \cdot \{B[1 + \exp(-2Bz)] - A(1 - \exp(-2Bz))\}\end{aligned} \tag{6.36}$$

Taking the limit as $z \to +\infty$, since $C(x, y, +\infty, t) = 0$, one obtains

$$\frac{\mathrm{d}\bar{C}^{xy}(w,\theta,0,s)}{\mathrm{d}z} = \frac{u}{D_z} \cdot \bar{g}^{xy} + (A-B) \cdot \bar{g}^{xy} \tag{6.37}$$

Inserting Eq. (6.37) into Eq. (6.35) yields

$$\bar{C}^{xy} = \bar{g}^{xy} \cdot \exp(-Az) \cdot \exp(-Bz) \tag{6.38}$$

The third term appearing on the right-hand side of Eq. (6.38) may be expressed in the following form:

$$h(x, y, z, s) = \exp(-Bz) = \exp\left[-\frac{z}{\sqrt{D_z(s+k_\mathrm{r})}}(s^2 + s(k_\mathrm{r} + E) + k_\mathrm{r} E - k_\mathrm{r} \cdot k_\mathrm{d})^{\frac{1}{2}}\right] \tag{6.39}$$

where $E = D_x w^2 + D_y \phi^2 + D_z A^2 + k_d$.

Furthermore, one obtains

$$L^{-1}[\frac{1}{s+k_r}\overline{f}_0(s+k_r-\frac{\alpha}{s+k_r})] = \exp^{-k_r t}\int_0^t I_0[2(\alpha\eta(t-\eta))^{1/2}]f_0(\eta)d\eta \tag{6.40}$$

where L^{-1} denotes the inverse Laplace operator, $\overline{f}_0(s)$ represents the Laplace transform of an arbitrary function $f_0(t)$, I_0 refers to the modified Bessel function of the first kind of order zero, and α denotes an arbitrary constant.

The $f_0(t)$ is expressed as

$$f_0(t) = \frac{\alpha_1}{(4\pi t^3)^{1/2}}\exp(-\frac{\alpha_1^{\ 2}}{4t} - \alpha_2 t) \tag{6.41}$$

Accordingly, taking the Laplace transform of Eq. (6.41) in the time domain gives the following result:

$$\overline{f}_0(s) = \exp\left[-\alpha_1(s+\alpha_2)^{1/2}\right] \tag{6.42}$$

where α_1 and α_2 represent arbitrary constants.

It follows from Eq. (6.42) that

$$\overline{f}_0(s+k_r-\frac{\alpha}{s+k_r}) = \exp\{-\frac{\alpha_1}{(s+k_r)^{1/2}}[s^2+s(2k_r+\alpha_2)+k_r(k_r+\alpha_2)-\alpha]^{1/2}\} \tag{6.43}$$

Let $\alpha = k_r k_d$, $\alpha_1 = z/\sqrt{D_z}$, and $\alpha_2 = E - k_r$. A comparison of Eqs. (6.39) and (6.43) yields

$$h(x,y,z,s) = \overline{f}_0(s+k_r-\frac{\alpha}{s+k_r}) \tag{6.44}$$

Therefore, Eq. (6.40) can be expressed as

$$L^{-1}(\frac{h}{s+k_r}) = \exp^{-k_r t}\int_0^t I_0[2(\alpha\eta(t-\eta))^{1/2}]f_0(\eta)\,d\eta \tag{6.45}$$

Assume $R(w,\phi,z,t) = L^{-1}[h/(s+k_r)]$. By applying the derivative property of the Laplace transform and using the condition $R(w,\phi,z,0) = 0$, we obtain $\partial R(w,\phi,z,t)/\partial t ==$ $L^{-1}[sh/(s+k_r)] = L^{-1}[s\overline{R}(x,y,z,s)]$.

Combining Eqs. (6.41) and (6.45) yields

$$R(w,\phi,z,t)=\exp^{-k_r t}\int_0^t I_0[2(\alpha\eta(t-\eta))^{1/2}]\cdot\frac{\alpha_1}{2\eta\sqrt{\pi\eta}}\cdot\exp[-\frac{\alpha_1^2}{4\eta}-\alpha_2\eta]d\eta \tag{6.46}$$

$$\frac{\partial R(w,\phi,z,t)}{\partial t}=\exp^{-k_r t}\int_0^t\{(\frac{\alpha\eta}{t-\eta})^{1/2}\cdot I_1[2(\alpha\eta(t-\eta))^{1/2}]-k_r I_0[2(\alpha\eta(t-\eta))^{1/2}]\}\cdot$$
$$\frac{\alpha_1}{2\sqrt{\pi\eta^3}}\exp[-\frac{\alpha_1^2}{4\eta}-\alpha_2\eta]d\eta+\exp^{-k_r t}\frac{\alpha_1}{2\sqrt{\pi t^3}}\exp[-\frac{\alpha_1^2}{4t}-\alpha_2 t] \tag{6.47}$$

where I_1 denotes the modified Bessel function of the first kind of order one.

For the sake of mathematical simplicity, the function $h(x,y,z,s)=\exp(-Bz)$ can be expressed as

$$h=\frac{k_r\cdot h}{s+k_r}+\frac{s\cdot h}{s+k_r} \tag{6.48}$$

Thereafter,

$$L^{-1}[\exp(-Bz)]=k_r\cdot R(w,\phi,z,t)+\frac{\partial R(w,\phi,z,t)}{\partial t} \tag{6.49}$$

Since $E=D_x\omega^2+D_y\phi^2+D_zA^2+k_d$, performing the inverse Fourier transform of $\exp(-Et)$ with respect to x yields

$$F_x^{-1}[\exp(-Et)]=\frac{1}{\sqrt{4\pi D_x t}}\cdot\exp[-\frac{x^2}{4D_x t}-(D_y\phi^2+D_zA^2+k_d)t] \tag{6.50}$$

In a similar manner, the inverse Fourier transform of Eq. (6.50) with respect to y can be expressed as follows:

$$F_y^{-1}\{F_x^{-1}[\exp(-Et)]\}=\frac{1}{4\pi t\sqrt{D_xD_y}}\cdot\exp(-\frac{x^2}{4D_x t}-\frac{y^2}{4D_y t}-\frac{u^2 t}{4D_z}-k_d t) \tag{6.51}$$

Considering Eq. (6.38) together with the convolution theorem, the solute concentration in the physical domain can be expressed as follows:

$$C(x,y,z,t)=\exp(\frac{uz}{2D_z})\cdot\int_0^t\int_{-\infty}^{+\infty}\int_{-\infty}^{+\infty}g(p,l,\tau)\cdot[k_r\cdot W(x-p,y-l,z,t-\tau)$$
$$+\frac{\partial W(x-p,y-l,z,t-\tau)}{\partial t}]\,d\,p\,d\,l\,d\tau \tag{6.52}$$

where p, l, and τ denote dummy integration variables, and

$$W(x,y,z,t)=\exp^{-k_r t}\int_0^t I_0[2(\alpha\eta(t-\eta))^{1/2}]\cdot\frac{z}{8\pi\eta^2\sqrt{\pi D_x D_y D_z \eta}}$$

$$\cdot\exp[-\frac{x^2}{4D_x\eta}-\frac{y^2}{4D_y\eta}-\frac{z^2}{4D_z\eta}-\frac{u^2\eta}{4D_z}+(k_r-k_d)\eta]\mathrm{d}\eta$$

$$\frac{\partial W(x,y,z,t)}{\partial t}=\exp^{-k_r t}\int_0^t\left\{(\frac{\alpha\eta}{t-\eta})^{1/2}I_1[2(\alpha\eta(t-\eta))^{1/2}]-k_r I_0[2(\alpha\eta(t-\eta))^{1/2}]\right\}$$

$$\cdot\frac{z}{8\pi\eta^2\sqrt{\pi D_x D_y D_z \eta}}\cdot\exp[-\frac{x^2}{4D_x\eta}-\frac{y^2}{4D_y\eta}-\frac{z^2}{4D_z\eta}-\frac{u^2\eta}{4D_z}+(k_r-k_d)\eta]\mathrm{d}\eta$$

$$+\exp^{-k_r t}\cdot\frac{z}{8\pi t^2\sqrt{\pi D_x D_y D_z t}}\cdot\exp[-\frac{x^2}{4D_x t}-\frac{y^2}{4D_y t}-\frac{z^2}{4D_z t}-\frac{u^2 t}{4D_z}+(k_r-k_d)t]$$

6.3.2 Contaminant Source Acting on the Local Surface of a Half-Space

6.3.2.1 Solution for a Point Contaminant Source

Consider an instantaneous point contaminant source with a solute concentration of $g(x,y,t)$, positioned at the coordinate (x', y', 0) on the surface of a semi-infinite domain (Figure 6.4). Under these conditions, the following relationship holds:

$$g(x,y,t)=I\cdot\delta(x-x')\delta(y-y')\,\delta(t-t') \tag{6.53}$$

where $I=M/u$ represents the intensity of the point source $[MTL^{-1}]$, where M denotes the total mass of the injected contaminant, $\delta(\cdot)$ denotes the Dirac delta function, the coordinates (x, y, z) specify the position of the observation point within the porous medium, while (x', y') correspond to the coordinates of the contaminant source plane (z = 0), t indicates the elapsed time, and t' denotes the time of contaminant injection.

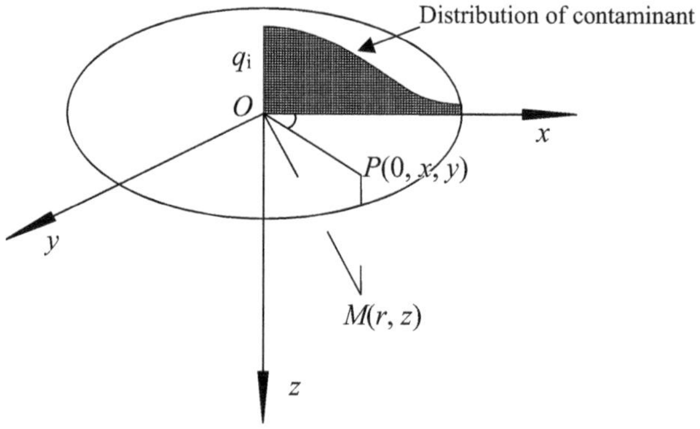

FIGURE 6.4 Physical model.

Substituting Eq. (6.53) into Eq. (6.52) within the interval $\int_{\alpha}^{\beta} f(t)\delta(t-t_0)\mathrm{d}t = f(t_0)$ ($a \le t_0 \le \beta$) leads to the derived solution that represents a point-contaminant source:

$$C(x,y,z,t) = \frac{I}{8\pi^{3/2}(D_x D_y D_z)^{1/2}} \cdot A_1(z,t-t') \\ \{\int_0^{t-t'} A_2(t-t')\cdot A_3(z,\eta)\cdot A_4(x,y,z,\eta)\mathrm{d}\eta + A_3(z,t-t')\cdot A_4(x,y,z,t-t')\} \tag{6.54}$$

where $A_1(z,t) = \exp(\frac{uz}{2D_z} - k_r t)$, $A_2(t) = (\frac{\alpha\eta}{t-\eta})^{1/2} \cdot I_1[2(\alpha\eta(t-\eta))^{1/2}]$, $A_3(z,t) = \frac{z}{t^{5/2}}$, and $A_4(x,y,z,t) = \exp[-\frac{(x-x')^2}{4D_x t} - \frac{(y-y')^2}{4D_y t} - \frac{z^2}{4D_z t} - \frac{u^2 t}{4D_z} + (k_r - k_d)t]$.

For the case where $I = 1$, the fundamental solution corresponding to an instantaneous point contaminant source can be derived from Eq. (6.54):

$$C_P(x,y,z,t) = \frac{1}{8\pi^{3/2}(D_x D_y D_z)^{1/2}} A_1(z,t-t') \\ \{\int_0^{t-t'} A_2(t-t')\cdot A_3(z,\eta)\cdot A_4(x,y,z,\eta)\mathrm{d}\eta + A_3(z,t-t')\cdot A_4(x,y,z,t-t')\} \tag{6.55}$$

6.3.2.2 Solution for a Localized Contaminant Source

Consider a localized contaminant region *A* characterized by a spatially and temporally varying intensity $q(x, y, t)$ $[ML^{-3}]$, defined as a function of time *t* and coordinates (*x*, *y*). This contaminant is introduced onto the surface of a semi-infinite, saturated porous medium at $t = 0$. Utilizing the source-function method proposed by Bai et al. (2015), the transport of the contaminant can be represented as the combined effect of numerous instantaneous, hypothetical point sources, each having an intensity given by $I = q(x, y, t)dxdydt$. The solute concentration can then be determined through integration of Eq. (6.55) over both spatial and temporal domains. Accordingly, the resultant expression describes the spatial distribution of solute concentration as follows:

$$C(x,y,z,t) = \int_0^t \iint_A C_P(x,y,z,t-\tau)\cdot q(x',y',\tau)\cdot \mathrm{d}x'\mathrm{d}y'\mathrm{d}\tau \tag{6.56}$$

A representative example of a circular contaminant zone with a time-dependent concentration $q(t)$ is illustrated in Figure 6.4. The coordinate system's origin *O* is positioned at the center of this contaminant source, which possesses a radius (*a*) and extends infinitely along the positive *z*-axis. It is assumed that the hydrodynamic dispersion coefficients are identical in the *x* and *y* directions, such that $D_x = D_y = D_r$, while the dispersion coefficient along the *z* direction is denoted as D_z. Under these conditions, the solute concentration distribution can be described within a cylindrical coordinate system (r, z) as follows:

$$C(r,z,t)=\frac{1}{8\pi D_r(\pi D_z)^{1/2}}\int_0^t q(\tau)A_1(z,t-\tau)$$
$$\cdot\{\int_{r'=0}^{a}\int_{\varphi'=0}^{2\pi}[\int_0^{t-\tau}A_2(t-\tau)\cdot A_3(z,\eta)\cdot A_4(r,z,\eta)\mathrm{d}\eta]\cdot r'\mathrm{d}\varphi'dr'$$
$$+\int_{r'=0}^{a}\int_{\varphi'=0}^{2\pi}A_3(z,t-\tau)\cdot A_4(r,z,t-\tau)\cdot r'\mathrm{d}\varphi'\mathrm{d}r'\}\mathrm{d}\tau \tag{6.57}$$

where $A_4(r,z,t)=\exp[-\frac{R^2}{4D_r t}-\frac{z^2}{4D_z t}-\frac{u^2 t}{4D_z}+(k_r-k_d)t]$, and $R=(r^2+r'^2-2rr'\cos\varphi')^{1/2}$.

Due to axial symmetry, integrating Eq. (6.57) with respect to φ gives the following result:

$$C(r,z,t)=\frac{1}{4D_r(\pi D_z)^{1/2}}\int_0^t q(\tau)A_1(z,t-\tau)$$
$$\left\{\begin{array}{l}\int_0^{t-\tau}[A_2(t-\tau)\cdot A_3(z,\eta)\cdot B_1(z,\eta)\cdot\int_{r'=0}^{a}B_2(r,z,\eta)\cdot r'\mathrm{d}r']\mathrm{d}\eta\\ +A_3(z,t-\tau)\cdot B_1(z,t-\tau)\cdot\int_{r'=0}^{a}B_2(r,z,t-\tau)\cdot r'\mathrm{d}r'\end{array}\right\}\mathrm{d}\tau \tag{6.58}$$

where $B_1(z,t)=\exp[-\frac{z^2}{4D_z t}-\frac{u^2 t}{4D_z}+(k_r-k_d)t]$ and $B_2(r,z,t)=\exp(-\frac{r^2+r'^2}{4D_r t})\cdot I_0(\frac{rr'}{2D_r t})$.

When $a=\infty$, Eq. (6.58) simplifies to a one-dimensional formulation. In the case where the contaminant source maintains a constant intensity (i.e., $q(x, y, t) = q$), the corresponding solute concentration field within the porous medium can be readily derived from Eq. (6.58).

6.3.3 Example Analyses

6.3.3.1 Problem and Transport Parameters

The solutions are applicable to contaminant transport phenomena in saturated porous media influenced by contaminant sources of arbitrary geometry and spatiotemporally varying concentrations. For clarity of illustration, this section focuses on two representative transport scenarios within a semi-infinite medium subjected to a circular contaminant source whose intensity changes with time, including exponential variation and harmonic variation. The corresponding multiple integrations are performed using Mathematica 9 software (Wolfram Research).

Selecting appropriate parameters for contaminant transport analysis is often a challenging task. Nevertheless, these parameters can be estimated by directly fitting the available experimental observations, such as results from one-dimensional column tests. It should be noted that the computed outcomes are highly sensitive to the chosen parameter values. Here, the contaminant transport parameters adopted for calculation are as follows: $k_r=0.001\,\mathrm{s}^{-1}$, $k_d=0.02\,\mathrm{s}^{-1}$, $u=0.2$ cm/s, $D_z=1\ \mathrm{cm}^2/\mathrm{s}$, and $D_r=0.1\ \mathrm{cm}^2/\mathrm{s}$.

6.3.3.2 *Transport Process under a Constant Contaminant Source*

In this scenario, a circular contaminant source is located on the surface of a semi-infinite porous medium. The contaminant concentration is assumed to decay exponentially with respect to time *t*. The spatial variation of the contaminant concentration along the radial direction *r* (Figure 6.4) can be described by the following function:

$$q(r,t) = q_i \exp(-\beta t).[\frac{1}{2}(1+\cos\frac{k\pi r}{a})] \tag{6.59}$$

where q_i represents the initial concentration of the contaminant source, β denotes the degradation coefficient, and k is a geometric parameter characterizing the shape of the contaminant source.

When $\beta = 0$ and $k = 0$, the expression becomes $q(r,t) = q_i$, and Eq. (6.59) simplifies to represent the case of a constant contaminant source. For analytical convenience, a dimensionless concentration denoted as C/C_0 is introduced, where C_0 refers to the initial concentration applied on the surface of the semi-infinite porous medium (i.e., $C_0 = q_i$).

Figure 6.5 illustrates the temporal and spatial evolution of contaminant concentration for the case of $q(r,t) = q_i$ and a = 10 cm. The results demonstrate that the dimensionless concentration C/C_0 gradually varies with both depth and lateral distance as time progresses (from t = 20 s to 200 s). Owing to the presence of convective flow in the depth direction, the contaminant predominantly migrates downward, and the transport velocity in this direction is considerably higher than that in the horizontal direction. After a sufficiently long period (e.g., t = 200 s), the system approaches a steady state, and the spatial concentration distribution remains essentially unchanged.

Figure 6.6 presents the variation of contaminant concentration with time at specific depths (z = 10, 20, and 30 cm) along the central axis of the contaminant source (r = 0) for different radii (a = 2 cm and 100 cm). The results indicate that, at each depth, the contaminant concentration gradually increases over time until a steady state is achieved, while the velocity of transport decreases with increasing depth. Additionally, the steady-state value of C/C_0 rises as the radius of the contaminant source becomes larger. For instance,

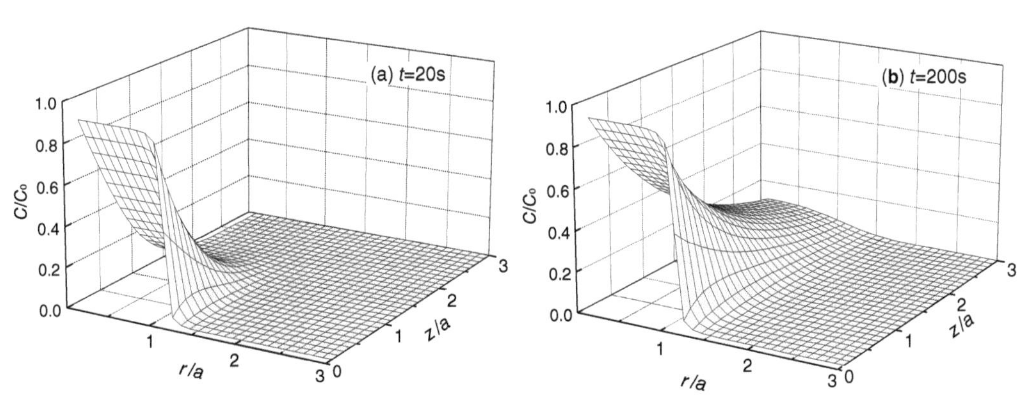

FIGURE 6.5 Spatial distribution of the contaminant. (a) t = 20 s. (b) t = 200 s.

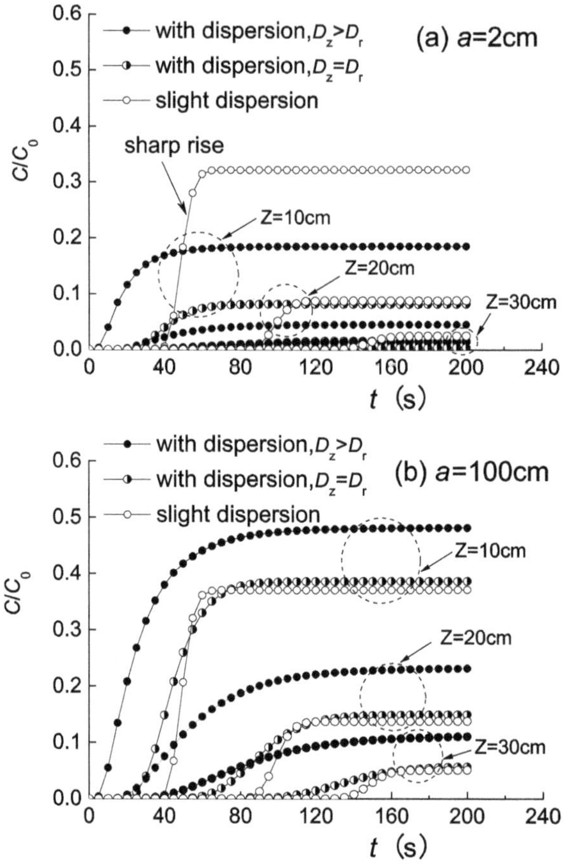

FIGURE 6.6 Hydrodynamic dispersion influence. (a) a = 2 cm. (b) a = 100 cm.

at a depth of z = 10 cm, the steady-state concentrations C/C_0 are 0.185 and 0.482 for a = 2 cm and 100 cm, respectively. Consequently, the penetration depth of the contaminant increases with source size. To further describe this behavior, the influence depth z_0 is defined as the depth beyond which the normalized concentration C/C_0 is less than 0.001. This implies that the contaminant impact at depths greater than z_0 is negligible, making z_0 a useful indicator of the pollution depth. The computed results show that when a = 2 cm, and 100 cm, the corresponding influence depths are $z_0 = 58$, and 98 cm (i.e., $z_0/a = 29$, and 0.98). These findings suggest that the influence depth approaches a constant value once a exceeds approximately 10 cm.

In the case of one-dimensional water flow, a distinct difference exists between the hydrodynamic dispersion in the vertical and horizontal directions, which is closely associated with the seepage velocity of the flow. The results indicate that when the hydrodynamic dispersion is sufficiently small to be considered negligible (i.e., $D_z = D_r = 0.01$ cm^2/s for slight dispersion; Figure 6.6), the transport process is dominated primarily by convective flow. Consequently, the transport velocity of the contaminant becomes slower compared with cases involving larger dispersion coefficients (i.e., $D_z = 1\ \text{cm}^2/\text{s}$ and $D_r = 0.1\ \text{cm}^2/\text{s}$). Under these conditions, the solute concentration rises rapidly from $C/C_0 = 0$ to a steady-state value within a relatively short period of time.

The steady-state contaminant concentrations observed near the surface of a porous medium (e.g., z = 10 and 20 cm) are found to be higher under conditions of slight dispersion than under strong dispersion when the source radius is relatively small (e.g., a = 2 cm; Figure 6.6a). In contrast, the opposite trend is evident when the contaminant source has a larger radius (e.g., a = 100 cm). This phenomenon occurs because dispersion enhances both vertical transport and horizontal diffusion, leading to a rapid increase in contaminant concentration during the early stage of transport. Nevertheless, for sources with a small radius, the steady-state concentration ultimately becomes lower than that obtained in the absence of dispersion, as lateral diffusion causes part of the solute to spread outward (e.g., a = 2 cm; Figure 6.6a). As the radius of the contaminant source increases, the horizontal diffusion effect weakens due to the growing dominance of the one-dimensional flow behavior, while the vertical dispersion becomes the controlling factor, resulting in greater solute accumulation within the medium.

However, when the vertical hydrodynamic dispersion coefficient is equal to the horizontal one (i.e., $D_z = D_r = 0.1\ \text{cm}^2/\text{s}$), the contaminant concentration during transport becomes lower than that observed in the case where $D_z = 1\ \text{cm}^2/\text{s}$ (Figure 6.6, where $D_z > D_r$). This reduction arises from the weakened vertical dispersion effect, which diminishes the contaminant's downward transport capability.

6.3.3.3 Transport Processes under an Exponentially Degenerated Contaminant Source

In this section, the value of β in Eq. (6.59) is set to 0.003 s^{-1}, and two scenarios are analyzed. In the first case, the contaminant source exhibits a uniform distribution along the radial direction (k = 0, case 1), where $q(r,t) = q_i \cdot \exp(-\beta t)$. In the second case, the contaminant source is nonuniformly distributed along the radius (k = 1; Case 2), defined by $q(0,t) = q_i \cdot \exp(-\beta t)$ and $q(a, t)$ = 0, as illustrated in Figure 6.4. The second condition frequently occurs in engineering applications, where the maximum contaminant concentration appears at the center of the source and gradually decreases to zero toward the outer boundary.

For Case 1, where k = 0 in Eq. (6.59), the contaminant concentration along the z direction attains its maximum within a relatively short time interval (e.g., from t = 30 s to t = 60 s; Figure 6.7a for a = 10 cm and z = 10, 20, 30 cm), after which it gradually declines with time. This behavior arises because the contaminant introduced into the porous medium migrates downward and outward under the combined influence of convective water flow, molecular diffusion, mechanical dispersion, and attachment on the solid matrix. Consequently, the contaminant concentration at a particular depth initially increases until reaching a peak value and then progressively decreases toward zero as the source intensity diminishes exponentially. With increasing depth (e.g., z = 10 cm to 20 cm to 30 cm), the time required to reach the maximum concentration becomes longer, and the peak value becomes smaller. Furthermore, the peak concentration decreases rapidly as the horizontal

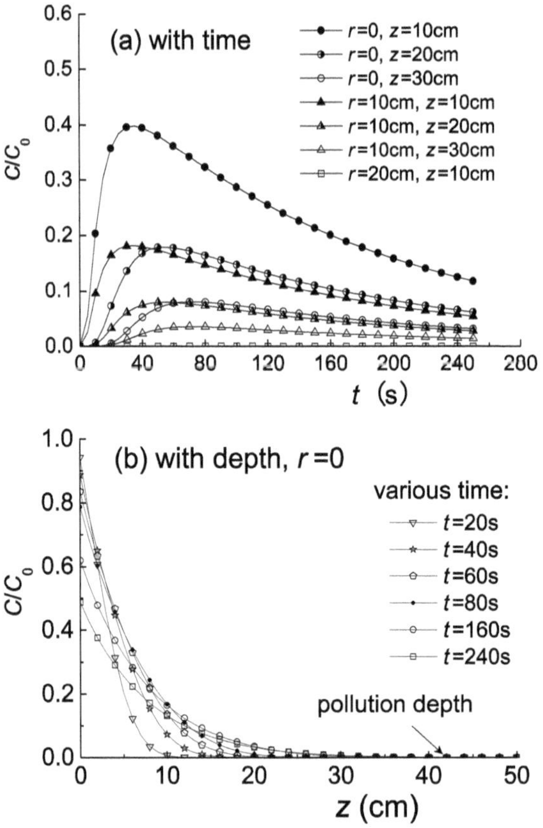

FIGURE 6.7 Transport processes under an exponentially degenerated contaminant source. (a) With time. (b) With depth, $r = 0$.

distance from the center of the contaminant source increases. This trend is particularly noticeable near the surface of the porous medium, where the concentration is less than $0.0001\,C_0$ (e.g., $r = 20$ cm and $z = 10$ cm; Figure 6.7a).

Figure 6.7b illustrates the variation of contaminant concentration along the depth direction ($r = 0$) in a porous medium influenced by an exponentially decaying contaminant source with a radius of $a = 10$ cm. The concentration near the surface of the medium first increases and then gradually decreases as time progresses. Meanwhile, the contaminant continues to migrate downward and eventually approaches zero with increasing depth. The maximum pollution depth is approximately $z_0 = 42$ cm, which corresponds to the condition $C/C_0 < 0.001$ when $z_0/a = 4.2$.

Figure 6.8 compares the temporal evolution of contaminant concentrations in two conditions: a uniform source distribution (Case 1, where $k = 0$ in Eq. (6.59), $a = 10$ cm, $r = 0$) and a nonuniform source distribution (Case 2, where $k = 1$). From Figure 6.8b, with increasing time ($t = 20$ s to 50 s to 200 s to 400 s) at a depth of $z = 10$ cm, both the concentration magnitude and the spatial extent of the contaminant are larger in Case 1 than in Case 2. A noticeable diffusive effect remains in the horizontal direction even at positions

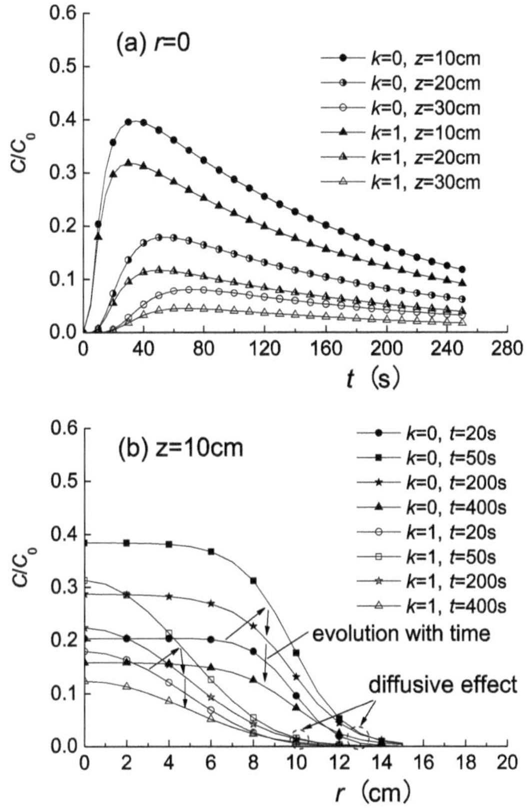

FIGURE 6.8 Comparisons of transport processes for uniform and nonuniform contaminant sources. (a) $r = 0$. (b) $z = 10$ cm.

distant from the center of the contaminant source, although the solute concentration in those outer regions is considerably reduced.

6.3.3.4 Transport Processes under a Harmonic Contaminant Source

Consider a harmonic contaminant source distributed over the circular surface of a semi-infinite medium (Figure 6.9). The concentration of the contaminant can be represented by the following expression:

$$q(t) = q_a + q_0 \cos(\omega t + \varphi) \tag{6.60}$$

where q_a denotes the mean intensity of the periodic contaminant source, q_0 represents its half-amplitude, and $\omega = 2\pi / Z$ is the angular frequency, where Z corresponds to the period.

For simplicity, the parameters in this section are taken as $q_a = q_0$, $Z = 50$ s, and $\varphi = 0$. The normalized concentration within a porous medium is defined as $C/C_0 = C/(2q_a)$. When $q_0 = 0$, Eq. (6.60) simplifies to represent the case of a constant contaminant source.

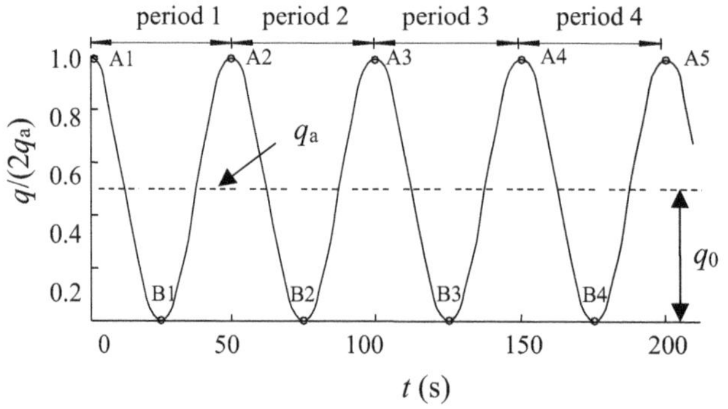

FIGURE 6.9 Variation of a cyclic contaminant source.

Figure 6.10 shows that the contaminant concentration at specified depths (e.g., z = 10, 20, and 30 cm) undergoes periodic oscillations as a result of the cyclic contaminant source (a = 10 cm) applied to the surface of the porous medium, eventually reaching a cyclic quasi-steady state. The phase of the concentration response lags behind that of the contaminant source. For instance, the maximum contaminant source concentration in periods 1–5 (denoted as A1, A2, A3, A4, and A5 in Figure 6.9) occurs at t = 0, 50, 100, 150, and 200 s, respectively, while the peak concentrations at a depth of z = 10 cm appear at t = 23, 68, 118, 168, and 218 s. The time intervals between successive peaks are 23, 45, 50, 50, and 50 s, respectively, demonstrating that the oscillation period of the concentration field matches that of the contaminant source (Z = 50 s). Moreover, the time required for C/C_0 to transition from a non–steady state to a quasi-steady state increases with depth (i.e., z = 10 cm to 20 cm to 30 cm), while both the amplitude of concentration fluctuations and the phase angle decrease as the depth increases.

Figure 6.10 also demonstrates that, at the initial stage, the contaminant concentration in a porous medium generated by a constant contaminant source ($q_0 = 0$) is generally higher than the average concentration produced by a cyclic contaminant source. A comparison between Figure 6.10a and b reveals that the time required for C/C_0 to progress from a non–steady state to a quasi–steady state increases with radial distance, while the amplitude of concentration fluctuations at a given depth decreases correspondingly. For instance, at r = 20 cm and z = 10 cm (Figure 6.10c), the value of C/C_0 is approximately 0.0002, which may be regarded as a nonpollution depth since $C/C_0 < 0.001$.

Figure 6.11 presents the spatial distribution of contaminant concentration in a porous medium influenced by a cyclic contaminant source at several representative phase points. The results indicate that the concentrations at phase points B1, B2, and B3 remain in a non–steady state. As the number of cycles increases, the contaminant gradually penetrates deeper into the porous medium, as observed from the sequence of valley points (B1 to B2 to B3 to B4) and peak points (A2 to A3 to A4 to A5). Eventually, the system approaches a quasi-steady condition, where cyclic fluctuations become nearly identical at the same phase point (e.g., A4 and A5, corresponding to t = 150 s and t = 200 s, respectively; Figure 6.11).

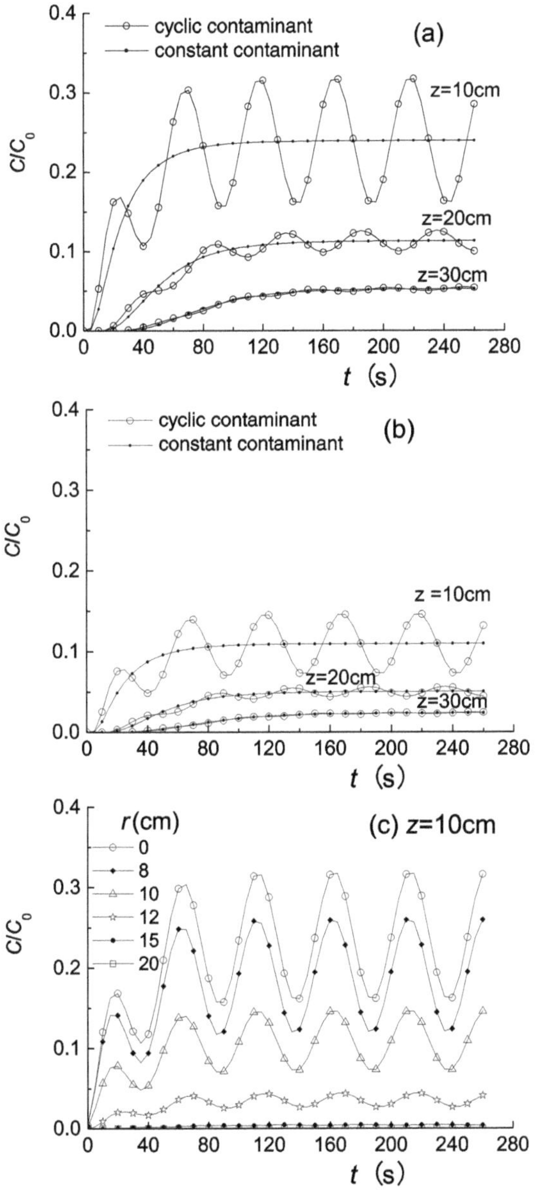

FIGURE 6.10 Transport processes under a cyclic contaminant source. (a) r = 0. (b) r = 10 cm. (c) z = 10 cm.

The differences between the normalized concentrations C/C_0 at these stages are less than 0.003. The maximum influence depth in the quasi-steady state is approximately $z_0 = 76$ cm ($z_0/a = 7.6$ with $C/C_0 < 0.001$), which is smaller than that obtained under the constant contaminant source condition ($z_0 = 92$ cm) but greater than that corresponding to the exponentially decaying source ($z_0 = 42$ cm; Figure 6.7b).

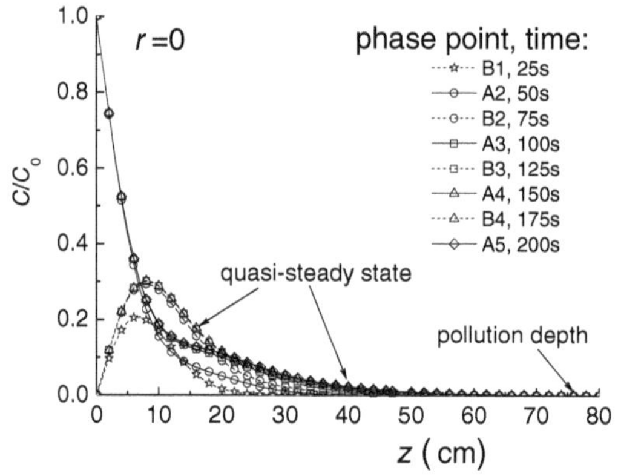

FIGURE 6.11 Distributions of contaminant concentration with depth.

Figure 6.12 demonstrates that when the hydrodynamic dispersion is relatively weak (i.e., $D_z = D_r = 0.01\ \text{cm}^2/\text{s}$), the transport velocity of the contaminant decreases noticeably, implying that a longer period is required for the contaminant concentration to reach a steady state. Moreover, the phase of the cyclic concentration peaks exhibits an observable delay, and the amplitudes of the steady-state peaks increase significantly. For instance, the first peak appears at t = 25 s under strong dispersion and at t = 55 s under weak dispersion, with corresponding concentrations of $C/C_0 = 0.168$ and 0.255, respectively. As the depth increases, the phase lag caused by the reduction in dispersion becomes more pronounced. At z = 20 cm, the first peak occurs at t = 105 s ($C/C_0 = 0.078$), which is considerably later than that at z = 10 cm (i.e., t = 40 s, $C/C_0 = 0.046$). This behavior highlights the influence of hydrodynamic dispersion on the contaminant transport process. In addition, with increasing radial distance from the center of the contaminant source, the horizontal diffusion becomes the dominant mechanism controlling transport (e.g., r = 20 cm; Figure 6.12c). Consequently, the three-dimensional diffusion effect causes the contaminant concentration at greater depths to remain relatively higher, although its absolute value remains low, typically below $C/C_0 = 0.0005$ (as observed when comparing z = 10 cm and z = 20 cm).

6.4 COTRANSPORT MODEL OF HMs AND SPs

6.4.1 Governing Equations

The cotransport of HMs and SPs, as illustrated in Figure 6.13, can occur through three primary pathways. First, a portion of the HMs or SPs becomes deposited on the matrix of the porous medium. Second, some remain dissolved or suspended in the pore water and are transported along with the fluid flow. Third, part of the HMs is adsorbed onto SPs, after which these complexes may either migrate with the water flow or become deposited onto the solid matrix. The adsorption/deposition processes of both HMs and SPs are further influenced by the desorption/release processes of previously deposited suspended substances, as represented by processes 1–4 in Figure 6.13.

Therefore, the cotransport of HMs and SPs involves the simultaneous movement of three distinct species: HMs, SPs, and SPs coupled with HMs (denoted by MPs). When the two species possess comparable particle sizes, their mutual coupling effects can be

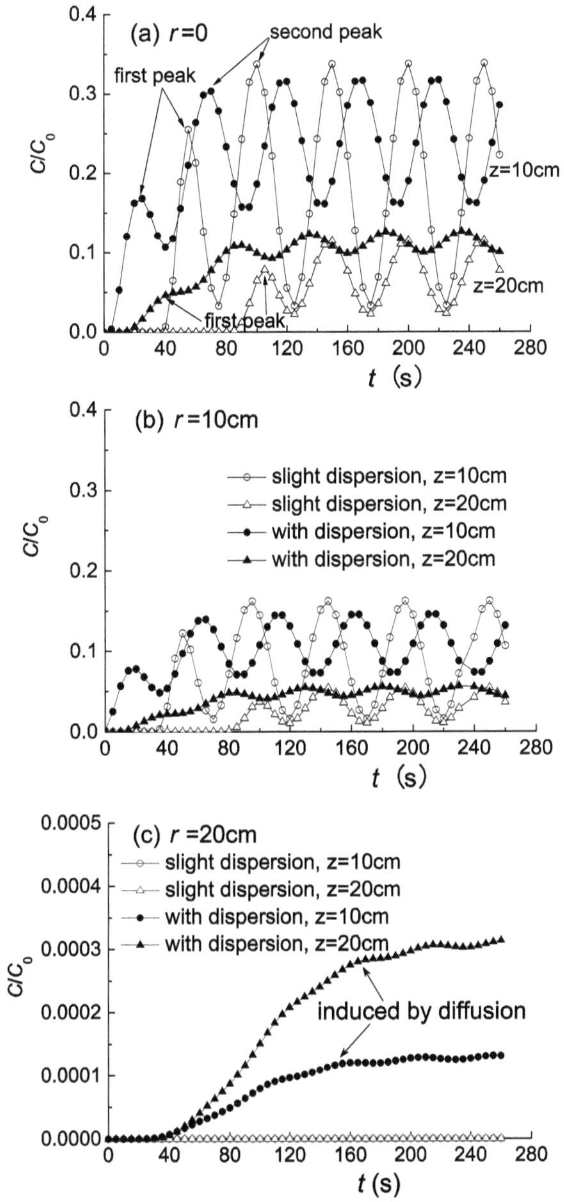

FIGURE 6.12 Diffusive effect on transport processes. (a) $r = 0$. (b) $r = 10$ cm. (c) $r = 20$ cm.

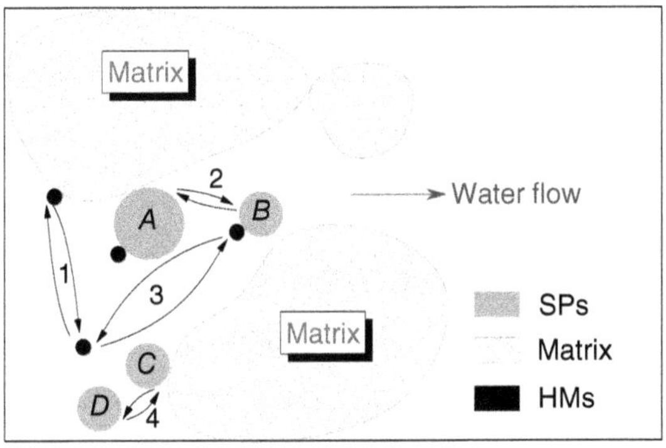

FIGURE 6.13 Cotransport of HMs and SPs.

considered equivalent. In contrast, when there are significant differences in particle size or dielectric properties, the influence of one constituent on the other may be negligible, while the reverse interaction becomes important. For example, in the cotransport of HMs and SPs, a portion of the HMs travels independently in the solution, whereas another portion is adsorbed onto the surface of SPs and migrates together with them (Figure 6.13). Based on these interactions, a new theoretical model for the cotransport of HMs and SPs is established, accounting for variations in both particle size and dielectric properties. Considering the strong coupling effect exerted by constituent j (e.g., SPs) on constituent i (e.g., HMs), the transport behavior of constituent i can be described by the convection–dispersion equation (Bai et al., 2021c; Wu et al., 2025):

$$\left\{\frac{\partial}{\partial t}\left[C_i(z,t)+\frac{\rho_s}{n}\sigma_i(z,t)\right]-D_i\frac{\partial^2 C_i(z,t)}{\partial z^2}\right\}+\left\{\frac{\partial}{\partial t}\left[C_j(z,t)\cdot C_{ij}(z,t)+\frac{\rho_s}{n}\sigma_j(z,t)\cdot\sigma_{ij}(z,t)\right]\right.$$
$$\left.-D_{ij}\frac{\partial^2}{\partial z^2}\left[C_j(z,t)\cdot C_{ij}(z,t)\right]\right\}=-u_i\frac{\partial}{\partial z}\left[C_i(z,t)+C_j(z,t)\cdot C_{ij}(z,t)\right] \qquad (6.61)$$

where z denotes the coordinate $[L]$; $\sigma_i(z,t)$ corresponds to the amount of species i deposited onto the solid matrix $[MM^{-1}]$; $\sigma_j(z,t)$ represents the concentration of constituent j deposited on the solid matrix $[MM^{-1}]$; $\sigma_{ij}(z,t)$ designates the concentration of constituent i absorbed on constituent j deposited on the solid matrix $[MM^{-1}]$; $C_i(z,t)$ represents the concentration of species i in the aqueous phase $[ML^{-3}]$; $C_j(z,t)$ denotes the concentration of constituent j in the aqueous phase $[ML^{-3}]$; $C_{ij}(z,t)$ refers to the concentration of constituent i absorbed on constituent j within the aqueous phase $[MM^{-1}]$; n denotes the porosity of the porous medium [–]; D_i and D_{ij} represent the hydrodynamic dispersion coefficients of species i and coupled species of i and j, respectively $[L^2T^{-1}]$; ρ_s denotes the density of the solid matrix $[ML^{-3}]$.

The dispersivity parameters of each species are expressed as $\alpha_{di}=D_i/u_0$ and $\alpha_{dij}=D_{ij}/u_0$, where u_0 denotes the average interstitial water velocity, defined by $u_0=v/n$. The first term on the left-hand side represents the transport process of constituent i within the aqueous phase, whereas the second term corresponds to the transport of constituent i adsorbed onto constituent j. The right-hand side of the equation accounts for the advection of constituent i carried by the flowing water and the movement of constituent i associated with constituent j. When both $C_{ij}(z,t)=0$ and $\sigma_{ij}(z,t)=0$, Eq. (6.61) simplifies to describe the transport of constituent i in the absence of any coupling effect from constituent j.

When SPs possess a mass much greater than that of HMs, the mass coupling influence exerted by HMs can be reasonably ignored. Nevertheless, because of the strong electrostatic attraction between HMs and SPs (e.g., SiO_2 particles) carrying opposite charges, HMs are readily adsorbed onto the SPs. Under these circumstances, the transport of SPs associated with HMs (i.e., MPs) becomes the primary transport mechanism (Figure 6.13;

particles A and B), whereas the movement of the free SPs (particles C and D) is insignificant. Consequently, the transport equation for SPs can be considerably simplified and regarded as identical to that describing the motion of individual SPs.

The adsorption of HMs onto SPs may alter the dielectric characteristics of the SPs surfaces, leading to corresponding variations in their dispersion coefficient, transport velocity, and attachment coefficient of the SPs. Since the mass of HMs is negligibly small compared with that of SPs, the overall mass of SPs carrying adsorbed HMs remains effectively unchanged. Furthermore, possible variations in the density (ρ_s) and porosity (n) of the porous medium are not considered in modeling the cotransport of HMs and SPs. Therefore, the transport of SPs associated with HMs (i.e., MPs) can be formulated based on Eq. (6.61) as follows:

$$\frac{\partial C_j(z,t)}{\partial t}+\frac{\rho_s}{n}\frac{\partial \sigma_j(z,t)}{\partial t}=D_{ij}\frac{\partial^2 C_j(z,t)}{\partial z^2}-u_{ij}\frac{\partial C_j(z,t)}{\partial z} \tag{6.62}$$

where u_{ij} refers to the transport velocity of MPs.

By replacing the subscript ij with j, the hydrodynamic equation governing the transport of SPs in the absence of HMs can be obtained, as expressed in Eq. (6.62). In this case, D_j denotes the hydrodynamic dispersion coefficient, and $\alpha_{dj}=D_j/u_0$ represents the dispersivity associated with SPs transport without HMs, while u_j denotes the mean interstitial velocity of SPs. The desorption (or detachment) of previously adsorbed HMs or MPs can markedly influence the attachment dynamics, particularly under variable-concentration injection conditions. The equation describing the detachment of HMs (process 1 in Figure 6.13) or MPs (process 2) from the solid matrix can therefore be derived using a linear kinetic model as follows:

$$\frac{\rho_s}{n}\frac{\partial \sigma_i(z,t)}{\partial t}=k_{di}\cdot C_i(z,t)-k_{ri}\cdot\frac{\rho_s}{n}\sigma_i(z,t) \tag{6.63}$$

where k_{di} and k_{ri} denote the attachment coefficient and detachment coefficient for HMs or MPs when the subscript i is replaced by j [T^{-1}].

It is evident that a complex coupling interaction exists both between mobile and immobile suspended substances (Figure 6.13, particles A and B) and between suspended substances and the solid matrix during the multiphase flow transport involving solid particles. The adsorption of HMs onto SPs that have settled on the solid matrix is largely governed by the concentration of MPs in the aqueous solution. When the detachment mechanism is considered, the concentration of HMs adsorbed on SPs already attached to the solid matrix can be described as follows (process 2 in Figure 6.13):

$$\frac{\partial}{\partial t}\left[\sigma_j(z,t)\cdot\sigma_{ij}(z,t)\right]=k_{d(j-ij)}\cdot\left[C_j(z,t)\cdot C_{ij}(z,t)\right]-k_{r(j-ij)}\cdot\frac{\rho_s}{n}\cdot\left[\sigma_j(z,t)\cdot\sigma_{ij}(z,t)\right] \tag{6.64}$$

where $k_{d(j-ij)}$ and $k_{r(j-ij)}$ denote the coefficients for attachment and detachment between MPs and the solid matrix, respectively $[T^{-1}]$.

HMs that are adsorbed onto SPs may be released back into the aqueous phase under the influence of hydrodynamic forces. The processes of suspended substance attachment and detachment within porous media are strongly affected by factors such as particle size, seepage velocity, and the maximum retention concentration. For the interaction between two suspended substances, a linear kinetic relationship is typically assumed to hold. Accordingly, the concentration of HMs adsorbed on SPs in the aqueous solution can be expressed as follows (process 3 in Figure 6.13):

$$\frac{\partial \sigma_{ij}(z,t)}{\partial t} = k_{dij} \cdot C_{ij}(z,t) - k_{rij} \cdot \sigma_{ij}(z,t) \tag{6.65}$$

where k_{dij} and k_{rij} represent the coefficients for attachment and detachment between HMs and SPs within the aqueous phase, respectively $[T^{-1}]$.

6.4.2 Determination of Transport Parameters

The hydrodynamic parameters relevant to the cotransport of HMs and SPs, including the dispersion coefficient, transport velocity, attachment coefficient, and detachment coefficient, are derived from experimental observations. The developed transport model can be used to fit the BTCs obtained in pulse experiments. Because it is often challenging to achieve accurate fitting when multiple parameters are adjusted simultaneously, the number of fitted parameters should be kept to a minimum to ensure reliable and distinct results. The method for estimating these parameters is presented below:

(i) The hydrodynamic dispersion coefficient $D_i (D_i = \alpha_{di} \cdot u_0)$, transport velocity u_i, attachment coefficient k_{di}, and detachment coefficient k_{ri} of HMs (i = HMs) are evaluated from Eqs. (6.61) and (6.63) based on the BTCs for single-species HMs transport. Because SPs are not present in this case, $C_{ij}(z,t) = 0$ and $\sigma_{ij}(z,t) = 0$ in Eq. (6.61) both are set to zero. For simplification, k_{ri} is assumed to be 0.1 k_{di}, since the detachment coefficient k_{ri} is generally much lower than the attachment coefficient k_{di}.

(ii) During cotransport, SPs are consistently associated with HMs, meaning free SPs do not exist independently. For SPs bound to HMs (i.e., MPs), their transport characteristics differ significantly from those of unbound SPs, primarily because the adsorbed HMs alter the surface charge of the SPs. The corresponding transport velocity u_{ij}, hydrodynamic dispersion coefficient $D_{ij} (D_{ij} = \alpha_{dij} \cdot u_0)$, attachment coefficient k_{dj}, and detachment coefficient k_{rj} for MPs can be determined from Eqs. (6.62) and (6.63) using the BTCs obtained from HMs–SPs cotransport experiments. As in the previous case, k_{rj} is taken as $0.1 k_{dj}$, which limits the number of fitting parameters to 3. For simplicity, the dispersion coefficient and transport velocity of MPs are assumed to be close to those of SPs, such as $D_{ij} = D_j$ and $u_{ij} = u_j$ in Eq. (6.62). Nevertheless, the experimental results indicate that the attachment coefficient of MPs differs markedly from that of SPs alone because of the electrostatic influence of HMs adsorbed on the particle surfaces.

(iii) In Eq. (6.64), because the attachment and detachment of HMs adsorbed on SPs are governed by the corresponding processes of MPs, it is assumed that $k_{\mathrm{d}(j-ij)} = k_{\mathrm{d}j}$ and $k_{\mathrm{r}(j-ij)} = 0.1k_{\mathrm{d}(j-ij)}$. Furthermore, the attachment coefficient $k_{\mathrm{d}ij}$ and detachment coefficient $k_{\mathrm{r}ij}$ describing the interaction between HMs and SPs (where ij = MP) can be obtained from isothermal-adsorption and desorption experiments. Since the ratio between these two coefficients is typically small, they are considered to be proportional (e.g., $k_{\mathrm{r}ij} = 0.1k_{\mathrm{d}ij}$).

6.4.3 Solutions and Verification

Table 6.2 presents the parameter values determined at a seepage velocity v = 0.087 cm/s and at temperatures of T = 20 °C, 40 °C, and 60 °C for Pb^{2+} solution. Table 6.2 also lists the corresponding adsorption and desorption coefficients for the HMs (i.e., Pb^{2+}) and SPs, under the assumptions that $k_{\mathrm{d}(j-ij)} = k_{\mathrm{d}j}$ and $k_{\mathrm{r}(j-ij)} = 0.1k_{\mathrm{d}(j-ij)}$.

A total of six unknown parameters need to be evaluated, namely, $C_i(z,t)$, $C_j(z,t)$, $\sigma_i(z,t)$, $\sigma_j(z,t)$, $C_{ij}(z,t)$, and $\sigma_{ij}(z,t)$. Specifically, $C_j(z,t)$ and $\sigma_j(z,t)$ for the SPs are obtained from Eqs. (6.62) and (6.63), while $C_{ij}(z,t)$ and $\sigma_{ij}(z,t)$ are calculated using Eqs. (6.64) and (6.65). The variables $C_i(z,t)$ and $\sigma_i(z,t)$ are estimated through Eqs. (6.61) and (6.63), respectively. The iterative computation is performed repeatedly until the desired convergence accuracy is achieved. The cotransport behavior of HMs and SPs is then simulated using the Pardiso solver implemented in COMSOL numerical software (COMSOL Co.).

Figure 6.14 presents the experimental and simulated BTCs for the cotransport of Pb^{2+} and SPs under single pulse-injection conditions, with D_{50}=2.8 μm and a seepage velocity of v = 0.087 cm/s. The simulation results indicate that the peak concentrations of Pb^{2+} decline as the temperature rises (T = 20 °C to 40 °C to 60 °C). The peak concentrations of SPs appear earlier than those of Pb^{2+}, suggesting that HMs require a longer period to reach a steady state. This behavior can be explained by the size exclusion effect influencing SPs and the molecular dispersion affecting HMs, as summarized in Table 6.2.

TABLE 6.2 Fitting Parameters at Multiple Temperatures

Temperaturee T (°C)	**HMs**				**SPs**				**MPs**	
	u_i/u_0	$a_{\mathrm{d}i}$ (cm)	$k_{\mathrm{d}i}(\mathrm{s}^{-1})$	R^2	u_j/u_0	$a_{\mathrm{d}ij}$ (cm)	$k_{\mathrm{d}j}(\mathrm{s}^{-1})$	R^2	$k_{\mathrm{d}(j-ij)}$	$k_{\mathrm{d}ij}$
20	0.52	0.73	0.013	0.96	0.71	0.69	0.007	0.97	0.007	0.01
40	0.53	0.75	0.012	0.96	0.73	0.68	0.008	0.96	0.008	0.01
60	0.58	0.56	0.013	0.97	0.78	0.41	0.010	0.98	0.010	0.01

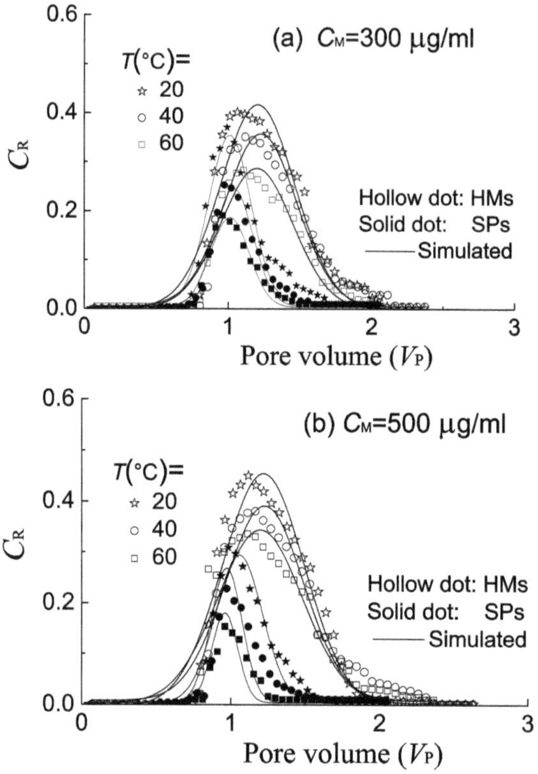

FIGURE 6.14 BTCs of HMs and SPs at various temperatures. (a) $C_M = 300$ μg/ml. (b) $C_M = 500$ μg/ml.

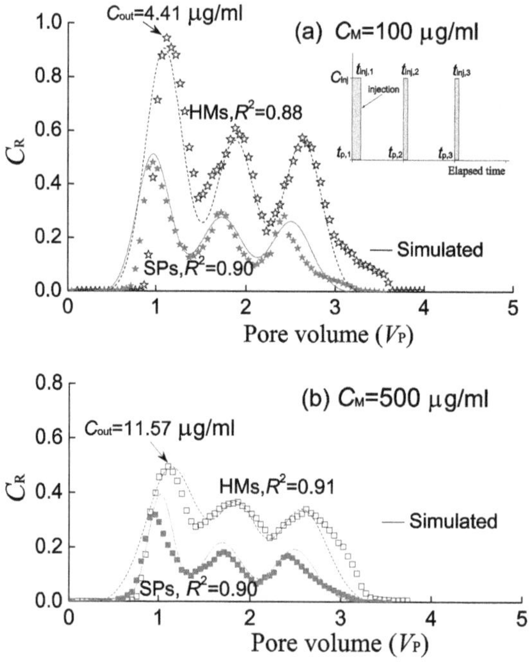

FIGURE 6.15 BTCs of HMs and SPs under repeated pulse-injections. (a) $C_M = 100$ μg/ml. (b) $C_M = 500$ μg/ml.

Figure 6.15 presents the corresponding results for repeated pulse-injections. The experimental data agree well with the simulated results, yielding $R^2 = 0.90-0.93$ in Figure 6.14 and $R^2 = 0.88-0.92$ in Figure 6.15. It is observed that increasing the input concentration of Pb^{2+} ($C_M = 100-500$ μg/ml) leads to higher Pb^{2+} concentration in the leachate ($C = 4.41-11.57$ μg/ml, $C_R = 0.94$ and 0.50), whereas the SPs concentration decreases correspondingly ($C = 0.0448$ to 0.0299 mg/ml, $C_R = 0.48$ and 0.32).

CHAPTER 7

Granular Thermodynamic Substance Flow Model

7.1 MAIN CONTENT OVERVIEW

A coupled multi-substance flow (CMF) model is established, which integrates the seepage, deformation, and suspended substance transport (including convection, diffusion, and attachment) within the framework of granular thermodynamics. The CMF model is particularly effective in considering the influence of temperature-driven forces. It evaluates various factors that impact the transport of suspended substances (e.g., suspended particles [SPs], heavy metal ions [HMs]) and integrates these factors into a comprehensive framework. With its broad applicability and flexibility, the model is well-suited for engineering applications, such as the remediation of contaminated soils considering both deformation and temperature effects, and the transport and recovery of pollutants in landfill environments.

For unsaturated soils, the CMF model is expanded to the coupled seepage–temperature conditions, including the transport of HMs and moisture migration. The model is validated using the experimental results under the transport of HMs, particularly in capturing the transition from an unsaturated state to a saturated state. Numerical calculations are performed to simulate the transport processes of moisture, water vapor, and HMs at different depths under constant/variable concentration conditions, with varying temperatures and seepage velocities.

7.2 COUPLED MULTI-SUBSTANCE FLOW MODEL WITH TEMPERATURE DRIVING EFFECT

7.2.1 Theory Framework

7.2.1.1 Mass, Momentum, and Energy Conservation

The transport system of porous media under external force is considered to consist of four phases: the solid matrix phase (S, the skeleton of porous media), the SPs phase (P, insoluble substances in water), the liquid water phase (W, saturated permeable water), and the HMs phase (H, soluble in water) (Figure 7.1). The model assumes the following: (i) mass exchange

DOI: 10.1201/9781003783749-7

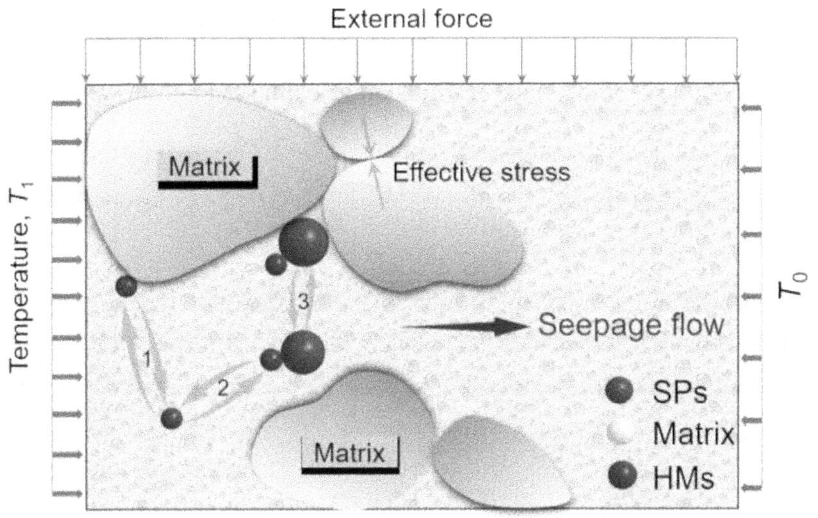

FIGURE 7.1 Mechanism of seepage, deformation, and substance transport.

occurs between HMs and the solid matrix (process 1), between HMs and SPs (process 2), and between SPs and the solid matrix (process 3); (ii) HMs moves with the same seepage velocity as the liquid water phase. The "double-entropy theory" (that is, including both thermodynamic entropy and granular entropy, along with thermodynamic temperature and granular temperature) is applied to describe the transport of suspended substances and the consolidation deformation of the porous media at the *meso*-level, while also considering the energy dissipation at the microscale caused by molecular thermal motion. Subsequent to this, the equations for mass, momentum, and energy conservation can be derived (Bai et al., 2021d, 2025c).

The mass conservation equation for each phase is formulated:

$$\frac{\partial}{\partial t}\rho^{\alpha} - D_{\alpha}\nabla^{2}\rho^{\alpha} + \nabla_{i}\left(\rho^{\alpha}v^{\alpha}\right) = m^{\alpha} \tag{7.1}$$

where ∇^2 denotes the Laplace operator, ∇ denotes the gradient operator; $\rho^{\alpha} = n^{\alpha}\rho_{\alpha}$, where n^{α}, v^{α}, ρ^{α}, and m^{α} denote the porosity of phase α, transport velocity, apparent density, and mass conversion rate (representing the attachment–detachment process or deposition–release process) between various phases $(\alpha = S, W, P, H)$, respectively; ρ_{α} indicates the intrinsic density of phase α; D_{α} indicates the diffusion coefficient of phase α, which applies specifically to $\alpha = P, H$.

The equation for the conservation of total momentum is expressed as follows:

$$\sum\nolimits_{\alpha=S,W,P,H}\frac{\partial}{\partial t}(\rho^{\alpha}v^{\alpha}) + \sum\nolimits_{\alpha=S,W,P,H}\nabla_{i}(\sigma^{\alpha}) + \sum\nolimits_{\alpha=P,W,H}\rho^{\alpha}(v_{i}^{\alpha} - v_{i}^{S})\nabla_{i}\left(v_{j}^{\alpha}\right) = \sum\nolimits_{\alpha=S,P,W,H}\rho^{\alpha}g \tag{7.2}$$

where σ^{α} indicates the Cauchy stress tensor of each phase; g indicates the acceleration of gravity.

The equation for total energy conservation can be expressed as follows:

$$\sum_{\alpha=S,W,P,H} \rho^{\alpha} \frac{\partial}{\partial t} e^{\alpha} = \sum_{\alpha=S,W,P,H} n^{\alpha} \sigma_{kj}^{\alpha} \psi_{jk}^{\alpha} + \nabla_k \left(w v_k + \sigma_{kj} v_j - Q_k \right) - \sum_{\alpha=W,P,H} \rho^{\alpha} v_k^{\alpha S} \nabla_k e_{\alpha} \tag{7.3}$$

where e^{α} denotes the specific energy of each phase; Q_k denotes the energy flow; v_k^{PS}, v_k^{WS}, and v_k^{HS} refer to the relative velocity of the SPs, water flow, and HMs to the solid matrix; ψ_{jk}^{α} denotes the deformation rate of each phase, defined as $\psi_{jk}^{\alpha} = -\left(\nabla_j v_k^{\alpha} + \nabla_k v_j^{\alpha} \right)/2$.

7.2.1.2 Thermodynamic Equation

The equation for entropy increase is written as

$$\sum_{\alpha=S,W,P,H} \rho^{\alpha} \frac{\partial}{\partial t} (\theta_{\alpha}) = \frac{R}{T} + \nabla_k f_k - \sum_{\alpha=W,P,H} \rho^{\alpha} v^{\alpha S} \nabla_k (\theta_{\alpha}) \tag{7.4}$$

The granular entropy increase is represented as

$$\rho^S d_t (\theta_G) = \frac{R_G}{T_G} - I_G \tag{7.5}$$

where R and R_G correspond to the generation rates of thermal entropy and granular entropy, respectively; f_k signifies the entropy conversion due to heat conduction; T denotes the absolute temperature (°C); T_G denotes the granular temperature; I_G denotes the decay rate of granular entropy; θ_G denotes the entropy capacity related to granular entropy; θ_{α} denotes the entropy capacity of each phase (i.e., the entropy per unit mass).

For the representative element that includes HMs and SPs, the independent state-variables are $A = \left\{ \rho^{\alpha}, m_i^{\alpha}, \varepsilon_{ij}^{e}, S, S_G \right\}$, with the corresponding conjugate set $B = \left\{ \mu_{ij}^{\alpha}, v_i^{\alpha}, \pi_{ij}, T, T_G \right\}$. Here, m_i^{α} corresponds to the kinetic energy of each phase, ε_{ij}^{e} represents the elastic strain, S and S_G denote the thermal entropy and granular entropy of the system, π_{ij} refers to the elastic stress of porous media, μ_{ij}^{α} is the chemical potential energy of each phase. The total differential form of thermodynamic energy is established by the state-variables and their conjugates:

$$\mathrm{d}w = \sum_{\alpha=S,W,P,H} \left(\mu_{ij}^{\alpha} \,\mathrm{d}\rho^{\alpha} + v_i^{\alpha} \,\mathrm{d}m_i^{\alpha} + T \,\mathrm{d}S^{\alpha} \right) + \pi_{ij} \,\mathrm{d}\varepsilon_{ij}^{e} + T_G \,\mathrm{d}S_G \tag{7.6}$$

Equation (7.6) shows that the thermodynamic potential energy includes various components: work done by the liquid-phase pressure, macroscopic average kinetic energy, heat flow, elastic potential energy, and the change in particle fluctuation or transport. By defining these energy forms, the functional expressions for both the total energy density and the free energy density are established. The thermodynamic identity is presented in terms of specific energy. From Eq. (7.6), the following expression can be derived:

$$\begin{aligned}\rho^{\alpha}\,\mathrm{d}e^{\alpha} = & \sum_{\alpha=S,W,P,H}\left[\mu_{ij}^{\alpha}\,\mathrm{d}\rho^{\alpha}+\frac{1}{2}v_{\alpha}^{2}\,\mathrm{d}\rho^{\alpha}+T\theta_{\alpha}\,\mathrm{d}\left(\rho^{\alpha}\right)-e^{\alpha}\,\mathrm{d}\rho^{\alpha}+T\rho^{\alpha}\,\mathrm{d}\left(\theta_{\alpha}\right)\right] \\ & +\pi_{ij}\,\mathrm{d}\varepsilon_{ij}^{\ e}+T_{\mathrm{G}}\rho^{S}\,\mathrm{d}\left(\theta_{\mathrm{G}}\right)+T_{\mathrm{G}}\theta_{\mathrm{G}}d\left(\rho^{S}\right)\end{aligned} \tag{7.7}$$

Next, to ensure the material system is in thermodynamic equilibrium within a sufficiently large spatial domain Ω filled with saturated rock and soil materials, the first-order variation of total entropy $\delta S=0$ and the second-order variation $\delta^{2}S<0$ must hold. Under this condition, the total entropy S reaches a maximum and is maintained at a constant total entropy state. The second-order variation for thermal entropy S with respect to time t is expressed as follows:

$$\begin{aligned}\delta S_{t} = & \iiint_{\Omega}\sum_{\alpha=S,W,P,H}\left(l_{m}^{S}-\frac{\mu_{ij}^{S}}{T}\right)\delta\rho^{S}-\frac{T_{\mathrm{G}}}{T}\delta S_{\mathrm{G}} \\ & +\sum\nolimits_{\alpha=S,W,P,H}\left(l_{i}^{k}-\frac{v_{i}^{\alpha}}{T}\right)\delta\rho^{\alpha}+\left(l_{w}+\frac{1}{T}\right)\delta w-\pi_{ij}\delta\varepsilon_{ij}^{\ e}\end{aligned} \tag{7.8}$$

where l_{m}^{S}, l_{i}^{k}, and l_{w} refer to the Lagrange multipliers corresponding to spatial and time invariance.

Based on the variation principle, Eq. (7.8) must hold for any variation of $\delta\rho^{\alpha}, \delta m^{\alpha}$, $\delta w, \delta\varepsilon_{ij}^{\ e}, \delta S_{\mathrm{G}}$, meaning the associated coefficients must be zero. This leads to further selection of the dissipative force and dissipative flow. In the coupled transport system of multi-substances in porous media, the dissipation force set C and the dissipation flow set D can be defined as $C=\left\{\pi_{ij}, T_{\mathrm{G}}, \nabla_{i}T, \psi_{ij}^{S}, \psi_{ij}^{W}, \psi_{ij}^{P}, \psi_{ij}^{H}, v_{i}^{W}-v_{i}^{S}, v_{i}^{P}-v_{i}^{S}, v_{i}^{H}-v_{i}^{S}\right\}$, $D=\left\{Y_{ij}, I_{\mathrm{G}}, q_{i}, \sigma_{ij}^{S}, \sigma_{ij}^{W}, \sigma_{ij}^{P}, \sigma_{ij}^{H}, \chi_{i}^{WS}, \chi_{i}^{PS}, \chi_{i}^{HS}\right\}$. For seepage transport problems, significant differences in seepage velocity cause the transport system to deviate substantially from equilibrium. Thus, adopting the difference in seepage velocity as the dissipative force is essential. Later, the relationship between the dissipative force and dissipative flow at the *meso*-level is established.

According to linear nonequilibrium theory, the dissipative flow can be represented as a linear function of the dissipative force, which gives $Y_{ij}=\eta_{ijkl}\cdot\pi_{ij}$, $I_{\mathrm{G}}=\gamma\cdot T_{\mathrm{G}}$, and $q_{i}=\kappa_{ik}/T\cdot\nabla_{i}T+\theta_{ik}^{\alpha S}\cdot\left(v_{i}^{\alpha}-v_{i}^{S}\right)$. In the equation for q_{i}, the first item represents heat flow under hydrostatic conditions, while the second term accounts for the coupling effect of isothermal heat flow. The viscous stress for each phase α ($\alpha=S,W,P,H$) is $\sigma_{ij}^{\alpha}=\lambda_{ijkl}^{\alpha}\cdot\psi_{ij}^{\alpha}$. The energy flow in phase α under isothermal conditions is expressed as $\chi_{i}^{\alpha S}=\beta_{ik}^{\alpha S}\left(v_{i}^{\alpha}-v_{i}^{S}\right)+\theta_{ik}^{\alpha S}\cdot\nabla_{i}T$, where $\eta_{ijkl}, \gamma, \frac{\kappa_{ik}}{T}, \lambda_{ijkl}^{S}, \lambda_{ijkl}^{W}, \lambda_{ijkl}^{P}, \lambda_{ijkl}^{H}, \beta_{ik}^{WS}, \beta_{ik}^{PS}$ and β_{ik}^{HS} are the migration coefficients for the primary dissipative flow; $\theta_{ik}^{WS}, \theta_{ik}^{PS}$, and θ_{ik}^{HS} refer to the thermo-osmotic coupling migration coefficients.

Within the framework of granular thermodynamics, the energy dissipation rate R is given by the product of dissipative force and dissipative flow:

$$\begin{aligned} R = {} & Y_{ij}\pi_{ij} + I_G T_G + q_i \nabla_i T + \sigma_{ij}^S \psi_{ij}^S + \sigma_{ij}^W \psi_{ij}^W \\ & + \sigma_{ij}^P \psi_{ij}^P + \sigma_{ij}^H \psi_{ij}^H + \chi_i^{WS}\left(v_i^W - v_i^S\right) \\ & + \chi_i^{PS}\left(v_i^P - v_i^S\right) + \chi_i^{HS}\left(v_i^H - v_i^S\right) \end{aligned} \tag{7.9}$$

For the transport of multi-substances, the generation of granular entropy can be written as

$$R_G = \sigma_{ij}^G \psi_{ij}^G + \sum\nolimits_{\alpha=S,W,P,H} f\left[U_\alpha, d_t\left(C_\alpha\right)\right] \tag{7.10}$$

where σ_{ij}^G refers to the dissipation force associated with porous medium deformation ψ_{ij}^G at the *meso*-level, and $\psi_{ij}^G = \psi_{ij}^S = \frac{\partial}{\partial t}\varepsilon_{ij}$. The term $f\left[U_\alpha, d_t\left(C_\alpha\right)\right] = T_G \cdot \Omega_\alpha \cdot d_t(C_\alpha)$ describes the energy dissipation resulting from the deviation of phase $\alpha\,(\alpha = P,H)$ from equilibrium resulting from the attachment or detachment of suspended substances. The dissipative flow for suspended substances is given by $\Omega_\alpha = \xi_\alpha U_\alpha$, where U_α represents the attachment concentration and ξ_α refers to the energy dissipation coefficient associated with the attachment or detachment process.

Substituting the conservation equations for mass, momentum, energy, thermodynamic entropy increase, and granular entropy increase into the thermodynamic identity (Eq. (7.7)), the resulting equation is

$$\begin{aligned} R + R_G = {} & \pi_{ij} d_t \varepsilon_{ij}^p + T_G I_G + f_k \nabla_k T + \psi_{jk}^S\left(\sigma_{kj}^S - \rho^S \mu^S \rho^S \delta_{ij} - \pi_{ij}\right) \\ & + \sum_{\alpha=W,P,H}\left[\psi_{jk}^\alpha\left(\sigma_{kj}^\alpha - \rho^\alpha \mu^\alpha \rho^\alpha \delta_{ij}\right)\right] \\ & + \sum_{\alpha=W,P,H} v_k^{\alpha S}\left[\rho^\alpha \mu^\alpha \nabla_i \rho^\alpha - \left(\nabla_k \sigma_{il}^\alpha - \rho^\alpha g + \rho^\alpha \nabla_k e_\alpha\right) + T\rho^\alpha \nabla_k \theta_\alpha\right] \\ & + \nabla_k\left(w v_k + \sigma_{kj} v_j - Q_k - f_k T\right) - \sum\nolimits_{\alpha=S,W,P,H}\left[\rho^\alpha \mu^\alpha (D_\alpha \nabla_i^2 \rho^\alpha + m^\alpha)\right] \end{aligned} \tag{7.11}$$

Combining Eqs. (7.9) and (7.10) leads to the equivalent of Eq. (7.11), from which the following conclusions can be drawn:

$$d_t \varepsilon_{ij}^p = Y_{ij} \tag{7.12}$$

$$q_i = f_k \tag{7.13}$$

$$w v_k + \sigma_{kj} v_j - Q_k = f_k T \tag{7.14}$$

$$\psi_{jk}^S\left(n^S \sigma_{kj}^S - \rho^S \mu^S \rho^S \delta_{ij} - \pi_{ij}\right) = \left(\sigma_{ij}^S + \sigma_{ij}^{vG}\right)\psi_{ij}^S \tag{7.15}$$

$$\psi_{jk}^\alpha\left(n^\alpha \sigma_{kj}^\alpha - \rho^\alpha \mu^\alpha \rho^\alpha \delta_{ij}\right) = \sigma_{ij}^\alpha \psi_{ij}^\alpha \qquad (\alpha = W,P,H) \tag{7.16}$$

$$\chi_i^{\alpha S} v^{\alpha S} = v_k^{\alpha S}\left(\rho^\alpha \mu^\alpha \nabla_i \rho^\alpha - \rho^\alpha \nabla_k e_\alpha - T\rho^\alpha \nabla_k \theta_\alpha\right) \quad (\alpha = W,P,H) \tag{7.17}$$

$$\sum\nolimits_{\alpha=S,W,P,H}\left[\rho^\alpha \mu^\alpha \left(D_\alpha \nabla_i^2 \rho^\alpha + m^\alpha\right)\right] = -\sum\nolimits_{\alpha=S,W,P,H} f\left(U_\alpha, d_t C_\alpha\right) \quad (\alpha = W,P,H) \tag{7.18}$$

where $\rho^\alpha \mu^\alpha \rho^\alpha = p^\alpha$ represents the pressure of each phase.

The pressure in both the SPs and HMs phases is the same as the pore water pressure. Since no mass exchange occurs between the HMs and water phases, the mass change rate of the water phase is zero, that is, $m^W = 0$. For the solid matrix phase on the left-hand side of Eq. (7.18), this leads to $-\rho^S \mu^S m^S = f\left[U_{HS}, d_t\left(C_{HS}\right)\right] + f\left[U_{PS}, d_t\left(C_{PS}\right)\right]$. For the SPs and HMs phases, the equations are $-\rho^P \mu^P m^P = f\left[U_{HP}, d_t\left(C_{HP}\right)\right] - f\left[U_{PS}, d_t\left(C_{PS}\right)\right]$ and $-\rho^H \mu^H m^H = f\left[U_H, d_t\left(C_H\right)\right]$. Thus, the mass change rate m^α of phase α is

$$m^\alpha = d_t\left(\rho_\alpha\right) = M_\alpha d_t\left(C_\alpha\right) \tag{7.19}$$

where $M_\alpha\left(\alpha = S,W,P,H\right)$ refers to the molar mass.

Combining Eqs. (7.15) and (7.10) with the migration coefficients, one can derive:

$$\begin{aligned}
\sigma_{ij}^S &= \lambda_{ijkl}^S d_t \varepsilon_{kl} = \left[\lambda_{sg} T_g \delta_{il} \delta_{jk} + \left(\lambda_{vg} - \frac{1}{3}\lambda_{sg}\right) T_g \delta_{ij} \delta_{lk}\right] d_t \varepsilon_{kl} \\
&= \lambda_{sg} T_g \delta_{il} \delta_{jk} d_t \varepsilon_{kl} + \left(\lambda_{vg} - \frac{1}{3}\lambda_{sg}\right) T_g \delta_{ij} \delta_{lk} d_t \varepsilon_{kl} \\
&= \lambda_{sg} T_g d_t \varepsilon_{ij} + \left(\lambda_{vg} - \frac{1}{3}\lambda_{sg}\right) T_g \delta_{ij} d_t \varepsilon_{kk}
\end{aligned} \tag{7.20}$$

Let the energy dissipation potential at the *meso*-level be defined as $w_G = \frac{1}{2} b\rho^S \left(T_G\right)^2$, and the granular entropy as $S_G = \frac{\partial w_G}{\partial T_G} = b\rho^S T_G$. By defining $T_{GG} = \lambda_{sg} T_G$ and combining it with Eq. (7.20), the equation for the conservation of granular temperature is obtained:

$$\begin{aligned}
d_t T_{GG} = {} & c_2 c_5 \frac{d_t \varepsilon_s d_t \varepsilon_s}{\rho^S} + c_3 c_5 \frac{d_t \varepsilon_v d_t \varepsilon_v}{\rho^S} \\
& + c_4 \frac{\sum\nolimits_{\alpha=W,P,H} f\left(\sigma_\alpha, d_t C_\alpha\right)}{\rho^S} - c_5 \frac{T_{GG}}{\rho^S}
\end{aligned} \tag{7.21}$$

As a result, the relationships between the stress field, temperature field, and the distribution of migrating substances are established.

7.2.1.3 Stress Field

The potential energy density function for porous media is given by

$$w_e = B_r \left[\frac{1}{\kappa + 2}\left(\varepsilon_v^e + \beta_T \Delta T\right)^{\kappa+2} + \chi\left(\varepsilon_v^e + \beta_T \Delta T\right)^\kappa \left(\varepsilon_s^e\right)^2\right] \tag{7.22}$$

where B_r is the stiffness of porous media; κ refers to the elastic potential energy parameter, typically set to $\kappa = 0.5$ (Bai et al., 2023); ε_v^e represents the elastic volumetric strain; β_T refers to the thermal expansion coefficient of solid matrix; ΔT represents the temperature difference; ε_s^e represents the elastic shear strain; χ represents the friction-related parameter.

The stiffness B_r is related to the porosity ratio e and is expressed as $B_r = B_0 \exp(-e/\lambda_0)$, where λ_0 is the hardening parameter and B_0 is the maximum stiffness. The elastic stress in the porous media, defined as $\pi_{ij} = \partial w_e / \partial \varepsilon_{ij}^e$, leads to

$$\pi_{ij} = B_r\left[\left(\varepsilon_v^e + \beta_T \Delta T\right)^{\kappa+1} \delta_{ij} + \kappa\chi\left(\varepsilon_v^e + \beta_T \Delta T\right)^{\kappa-1}\left(\varepsilon_s^e\right)^2 \delta_{ij} + 2\chi\left(\varepsilon_v^e + \beta_T \Delta T\right)^{\kappa} e_{ij}^e\right] \quad (7.23)$$

The elastic bulk modulus K_e and elastic shear modulus G_e are expressed as follows:

$$K_e = B_r\left[\left(\varepsilon_v^e + \beta_T \Delta T\right)^{\kappa} + \kappa\chi\left(\varepsilon_v^e + \beta_T \Delta T\right)^{\kappa-2}\left(\varepsilon_s^e\right)^2\right] \quad (7.24)$$

$$G_e = B_r \chi\left(\varepsilon_v^e + \beta_T \Delta T\right)^{\kappa} \quad (7.25)$$

From Eq. (7.23), the stress equation simplifies to

$$\pi_{ij} = K_e\left(\varepsilon_v^e + \beta_T \Delta T\right)\delta_{ij} + 2G_e e_{ij}^e \quad (7.26)$$

Combining the migration coefficients with the relation $Y_{ij} = \eta_{ijkl} \cdot \pi_{ij}$, one derives that the following:

$$d_t \varepsilon_{ij}^p = c_1 T_{GG}\left[\left(\varepsilon_v^e + \beta_T \Delta T\right)\right]\delta_{ij} + T_{GG} e_{ij}^e \quad (7.27)$$

For $i = j$, this represents the plastic volumetric strain: $d_t \varepsilon_{ij}^p = d_t \varepsilon_v^p = 3c_1 T_{GG}\left[\left(\varepsilon_v^e + \beta_T \Delta T\right)\right]$. For $i \neq j$, it corresponds to the plastic shear strain: $d_t \varepsilon_{ij}^p = d_t \varepsilon_s^p = T_{GG} \varepsilon_s^e$. Using Eq. (7.26), one can further derive an equation for the deviatoric stress:

$$q = \sqrt{\frac{3}{2} S_{ij} S_{ij}} = \sqrt{\frac{3}{2}\left(2G_e e_{ij}^e\right)\left(2G_e e_{ij}^e\right)} = \sqrt{6} B_r \eta\left(\varepsilon_v^e + \beta_T \Delta T\right)^{\kappa} \varepsilon_s^e \quad (7.28)$$

7.2.1.4 Temperature Field

The specific energies are defined as follows: $e^S = c_S T + \left[w_e\left(\rho_d, \varepsilon_{ij}^e, T\right) + w_G\left(\rho_d, T_G\right) / \rho^S\right)\right]$, $e^W = c_W T$, $e^P = c_P T$, $e^H = c_H T$ (Bai et al., 2023). Substituting these expressions for specific energies into the left-hand side of the energy conservation equation (i.e., Eq. (7.3)) gives

$$\begin{aligned} \sum_{\alpha=S,W,P,H} \rho^\alpha d_t e^\alpha &= \rho^S\left(c_S d_t T + \frac{1}{\rho^S} d_t w_e\right) + \sum_{\alpha=W,P,H} \rho^\alpha c_\alpha d_t T \\ &= \sum\nolimits_{\alpha=S,W,P,H}(\rho^S c_S d_t T) + \frac{\partial w_e}{\partial T} d_t T \end{aligned} \quad (7.29)$$

The right-hand side of Eq. (7.3) can be simplified by applying the thermodynamic identity (Eq. (7.7)) and the migration coefficients:

$$\begin{aligned}\nabla_k\left(wv_k+\sigma_{kj}v_j-Q_k\right)&=\nabla_k\left[\left(\kappa_{ik}\nabla_iT/T+\theta_{ik}^{WS}v^{WS}+\theta_{ik}^{PS}v^{PS}+\theta_{ik}^{HS}v^{HS}\right)\cdot T\right]\\&=\nabla_k\left(\kappa_{ik}\nabla_iT\right)+\nabla_k\left[\left(\theta_{ik}^{WS}v^{WS}+\theta_{ik}^{PS}v^{PS}+\theta_{ik}^{HS}v^{HS}\right)\cdot T\right]\end{aligned} \tag{7.30}$$

By combining Eqs. (7.29), (7.30), and (7.3), the temperature field equation is obtained:

$$\begin{aligned}\sum_{\alpha=S,W,P,H}\rho^\alpha c_\alpha d_tT+\frac{\partial w_e}{\partial T}d_tT&=\nabla_k\left[\left(\kappa_{ik}\nabla_iT\right)\right]+\nabla_k\left[\sum_{\alpha=W,P,H}\theta_{ik}^{\alpha S}v^{\alpha S}T\right]-c_W\rho^Wv_k^{WS}\nabla_k(T)\\&-c_P\rho^Pv_k^{PS}\nabla_k(T)-c_H\rho^Hv_k^{HS}\nabla_k(T)+\sum\nolimits_{\alpha=S,W,P,H}n^\alpha\sigma_{kj}^\alpha\psi_{jk}^\alpha\end{aligned} \tag{7.31}$$

7.2.1.5 Distribution Field of Transport Substances

The internal energy dissipation rate identity for a closed system, as described in Eqs. (7.7) and (7.11), leads to the following equation:

$$\rho^\alpha\mu^\alpha\nabla_i\rho^\alpha-\left(\nabla_k\sigma_{il}^\alpha+\rho^\alpha\nabla_ke^\alpha\right)+T\rho^\alpha\nabla_k\theta_\alpha=\chi_i^{\alpha S}\qquad(\alpha=W,P,H) \tag{7.32}$$

To consider the effect of the temperature gradient, the relationship between dissipation force $\left(v_i^W-v_i^S\right)$, $\left(v_i^P-v_i^S\right)$, $\left(v_i^H-v_i^S\right)$ and dissipation flow χ_i^{WS}, χ_i^{PS}, χ_i^{HS} can be expressed as

$$\chi_i^{\alpha S}=\theta_{ik}^{\alpha S}\cdot\nabla_kT+\beta_{ik}^{\alpha S}\left(v_i^\alpha-v_i^S\right)\qquad(\alpha=W,P,H) \tag{7.33}$$

The specific energies e^W, e^P, e^H are governed by the first law of thermodynamics, yielding

$$\nabla_ke^\alpha=T\nabla_k\theta_\alpha+\mu^\alpha\nabla_k\rho^\alpha\qquad(\alpha=W,P,H) \tag{7.34}$$

In Eq. (7.34), the first term represents the heat change, and the second term refers to volumetric work by pressure. Substituting Eqs. (7.33) and (7.34) into Eq. (7.32) results in

$$\theta_{ik}^{\alpha S}\cdot\nabla_kT+\beta_{ik}^{\alpha S}\left(v_i^\alpha-v_i^S\right)=-\nabla_k\sigma_{il}^\alpha\qquad(\alpha=W,P,H) \tag{7.35}$$

Next, the force and pore pressure are defined as follows (Bai et al., 2023):

$$\sigma_{kj}^W=\rho^W\mu^W\rho^W\delta_{ij}=p^W \tag{7.36}$$

Substituting Eq. (7.36) into Eq. (7.35) gives

$$\beta_{ik}^{\alpha S}\left(v_i^\alpha-v_i^S\right)=-\nabla_i\left(p^\alpha\right)-\theta_{ik}^{\alpha S}\cdot\nabla_kT\qquad(\alpha=W,P,H) \tag{7.37}$$

The migration coefficients β_{ik}^{WS}, β_{ik}^{PS}, β_{ik}^{HS} are linked to the permeability coefficient. These are assumed to be $\beta_{ik}^{WS}=\frac{u_W}{K\cdot k_W}\delta_{ik}$, $\beta_{ik}^{PS}=\frac{u_P}{K\cdot k_W}\delta_{ik}$, $\beta_{ik}^{HS}=\frac{u_H}{K\cdot k_W}\delta_{ik}$. Using this, Eq. (7.37) can be rewritten as follows:

$$v_i^{\alpha}-v_i^{S}=-\frac{K\cdot k_{\alpha}}{u_{\alpha}}\nabla_i\left(p^{\alpha}\right)-\frac{K\cdot k_{\alpha}}{u_{\alpha}}\theta_{ik}^{\alpha S}\cdot\nabla_k T \qquad (\alpha=W,P,H) \tag{7.38}$$

where k_{α} represents the relative permeability; u_{α} denotes the viscosity coefficients of each phase, with $u_W=u_H$; K represents the permeability coefficient (mD).

Additionally, define the deformation rate for phase α as $\psi_{ij}^{\alpha}=-\left(\nabla_i v_j^{\alpha}+\nabla_j v_i^{\alpha}\right)/2$. The time derivative of a substance with the solid phase as the reference is given by $d_t=\partial_t+v^S\nabla_i$. For the solid skeleton, one has $d_t\varepsilon_{ij}=\psi_{ij}^{S}-\varepsilon_{ik}\nabla_j v_k-\varepsilon_{jk}\nabla_i v_k$. Assuming small deformations, the last two terms can be neglected, and $d_t\varepsilon_{ij}=\psi_{ij}^{s}$. For the water phase, HMs, and SPs, Eq. (7.1) becomes

$$\frac{\partial}{\partial t}\rho^{\alpha}+\nabla_i(\rho^{\alpha}v^{\alpha})=d_t\rho^{\alpha}=-\nabla_i\rho^{\alpha}\left(v^{\alpha}-v^{S}\right)-\rho^{\alpha}d_t\psi_{kk}^{S}+D_{\alpha}\nabla^2\rho^{\alpha}+m^{\alpha} \qquad (\alpha=W,P,H) \tag{7.39}$$

The distribution field of suspended substances in porous media is obtained by inserting Eq. (7.38) into Eq. (7.39):

$$\begin{aligned} &d_t\left(\rho_{\alpha}\right)+\nabla_i\left\{\left(\rho_{\alpha}\right)\left[-\frac{K\cdot k_{\alpha}}{u_{\alpha}}\nabla_i\left(p^{\alpha}\right)-\frac{K\cdot k_{\alpha}}{u_{\alpha}}\theta_{ik}^{\alpha S}\nabla_k T\right]\right\} \\ &+\rho_{\alpha}d_t\psi_{kk}^{S}=D_{\alpha}\nabla^2\rho^{\alpha}+m^{\alpha} \quad (\alpha=W,P,H) \end{aligned} \tag{7.40}$$

7.2.2 Attachment/Detachment of Multi-Substance Transport

During the transport of multi-substances (Figure 7.1), water-soluble HMs, such as Pb^{2+}, flow at the same seepage velocity as liquid water (i.e., $v^L=v^H$). However, the interaction between water-insoluble SPs (e.g., nano- or micrometer-scale particles) and the porous medium matrix causes their seepage velocity to differ from that of liquid water (i.e., $v^L\neq v^P$).

The total mass of the HMs is significantly smaller compared to the mass of the porous medium matrix or SPs, so changes in the mass of the porous medium matrix or SPs due to the attachment and detachment processes of HMs (H) with the solid matrix (S) or SPs (P) can be neglected. However, the mass change of the HMs during these exchange processes must be considered. Specifically, the HMs mass exchange is represented as $m^{HS}=m^S$, $m^{HP}=m^P$, and $m^H=m^{HS}+m^{HP}$. Notably, SPs also attach to or detach from the solid matrix during transport, which reduces the mass of the SPs phase while increasing the mass of the porous medium matrix phase, resulting in a phase transition between the two (i.e., $m^P=m^S$).

For water-soluble HMs, an isothermal linear attachment–detachment model (Bai et al., 2019b) is used to describe the attachment and detachment processes between the HMs and the solid matrix ($\alpha = S$), and between the HMs and SPs ($\alpha = P$). The HMs mass exchange process is expressed as follows:

$$m^{\alpha H} = k_{\alpha \mathrm{d}} \dot{C}_H - k_{\alpha \mathrm{r}} \dot{C}_H \qquad (\alpha = S, P) \tag{7.41}$$

where $k_{\alpha \mathrm{r}}$ refers to the linear detachment coefficient; $k_{\alpha \mathrm{d}}$ $(\alpha = P, S)$ refers to the linear attachment coefficient.

The Bai model (Bai et al., 2025c) describes the attachment and detachment of SPs to the solid matrix. The mass exchange term (i.e., Eq. (7.1)) for the multi-substance is then given by

$$m^P = \rho^S \cdot \frac{\partial U_\alpha}{\partial t} = \rho^S \cdot \left[\dot{S}_\alpha + e^{-\lambda t} \left(\dot{S}_{\mathrm{r}\alpha} - \dot{S}_\alpha \right) + \lambda \cdot e^{-\lambda t} \left(S_\alpha - S_{\mathrm{r}\alpha} \right) \right] \tag{7.42}$$

where U_α denotes the amount of suspended substance deposited on the solid matrix under dynamic flow conditions, expressed as $U_\alpha = S_\alpha \left(1 - e^{-\lambda_\alpha t} \right) + S_{\mathrm{r}\alpha} e^{-\lambda_\alpha t}$; here, $S_{\mathrm{r}\alpha}$ denotes the equilibrium residual concentration, S_α denotes the equilibrium attachment concentration, and λ_α represents the reaction rate constant (see Chapter 3) between solid matrix and suspended substance.

The static nonlinear attachment–detachment model by Bai et al. (2019a) is commonly used to describe the attachment–detachment behavior of soluble ions, nanoscale colloids, and micrometer-scale SPs (i.e., Bai model). This model has been thoroughly validated (see Chapter 3). It provides the following expressions for the attachment and residual concentrations of suspended substances on the solid matrix:

$$U_\alpha = \frac{\kappa_{\mathrm{d}} C_\eta}{\beta_{\mathrm{d}}} \left(1 - e^{-\beta_{\mathrm{d}} \cdot \frac{C_\varsigma}{C_\eta}} \right) - \frac{\kappa_{\mathrm{r}} C_\varsigma}{\beta_{\mathrm{r}}} \left(e^{-\beta_{\mathrm{r}} \cdot \frac{C_P}{C_\varsigma}} - e^{-\beta_{\mathrm{r}}} \right) \tag{7.43}$$

$$U_{\mathrm{r}\alpha} = \frac{\kappa_{\mathrm{d}} C_\eta}{\beta_{\mathrm{d}}} \left(1 - e^{-\beta_{\mathrm{d}} \cdot \frac{C_\varsigma}{C_\eta}} \right) - \frac{\kappa_{\mathrm{r}} C_\varsigma}{\beta_{\mathrm{r}}} \left(1 - e^{-\beta_{\mathrm{r}}} \right) \tag{7.44}$$

where κ_{d} and β_{d} denote the transient equilibrium constant and attenuation factor for the attachment process, respectively; C_η refers to the characteristic concentration (usually set to 1); κ_{r} and β_{r} denote the transient equilibrium constant and attenuation factor for the detachment process; C_ς refers to the initial attachment concentration (being equal to C_{P}).

7.2.3 Substance Transport under External Force, Seepage, and Temperature

7.2.3.1 Calculation Example and Method

A two-dimensional axisymmetric finite element method (FEM) simulation was conducted to examine the feasibility of the developed CMF model for seepage transport under lateral

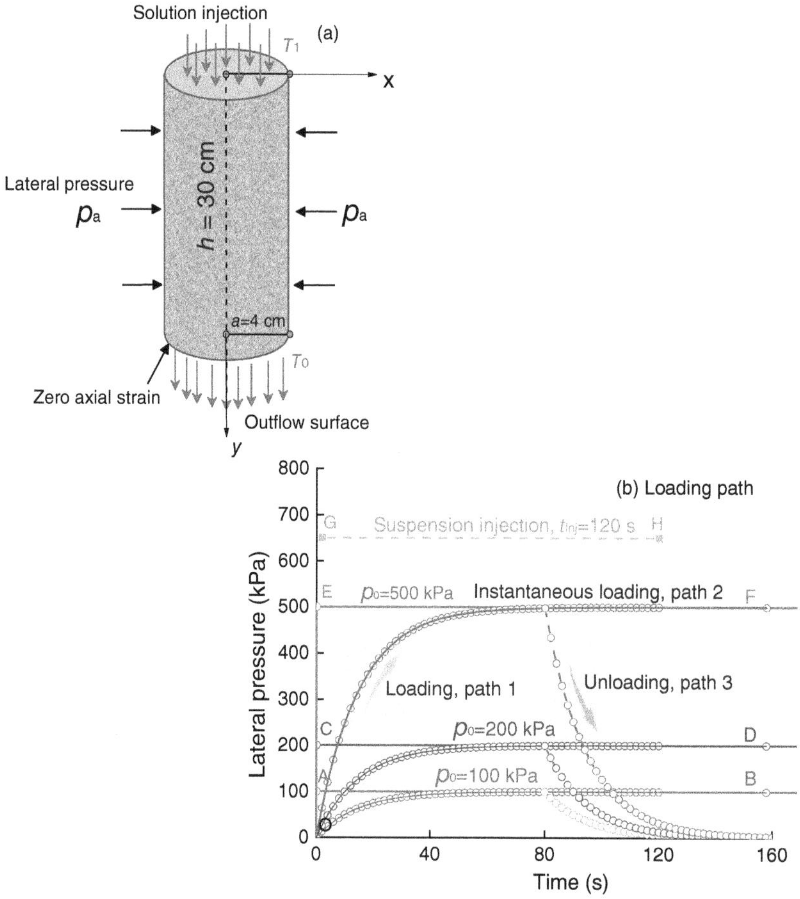

FIGURE 7.2 Transport calculation example. (a) Physical model. (b) Loading path.

(i.e., radial) force, as shown in Figure 7.2. The initial temperature was set at $T_0 = 20^{\circ}\mathrm{C}$. Seepage flow and temperature load (T_1) were applied vertically from the top to the bottom (T_0), with an external lateral force applied. The porous medium column (e.g., sand column) had a radius of $a = 4$ cm and a height of $h = 30$ cm, with an initial void ratio of $e = 0.715$. The porous medium particle density was $\rho_S = 2.65\times10^3\ \mathrm{kg/m^3}$.

The upper and lower surfaces of the sand column are fixed, with zero axial displacement, as shown in Figure 7.2a. Different lateral pressures are applied to the sidewall, allowing lateral deformation of the sand column. The CMF model accounts for the combined effects of seepage, external force, and temperature change over time. The sand column is assumed to have fully permeable conditions on the sides, meaning excess pore water pressure is zero under lateral pressure. Considering the loading rate effect (i.e., rate-dependent property), it is assumed that the lateral force on the sidewall increases exponentially with time, leading to varying lateral pressures (i.e., loading path 1, Figure 7.2b). The loading process is described by

$$p_a(t) = p_0\left[1 - \exp(-\omega t)\right] \tag{7.45}$$

where the load amplitudes are $p_0 = 100, 200, 500$ kPa, and the parameter $\omega = -0.00684\ s^{-1}$. When $\omega = 0$, Eq. (7.45) corresponds to an instant application of a constant load (path 2 in Figure 7.2b), where $p_a(t) = p_0$. The scenario of unloading after reaching a certain load value is also considered (unloading starts at $t_0 = 80\ s$, path 3), with the load change described by $p_a(t) = p_0 \exp[-\omega(t - t_0)]$ (Figure 7.2b). Along with the lateral force, a mixed solution of SPs and HMs is injected at the top of the sand column for a brief period ($t_{\text{inj}} = 120$ s). The SPs density is $\rho_P = 2.2 \times 10^3\ \text{kg/m}^3$. The initial concentrations are $C_{P0} = 2$ mg/ml for SPs and $C_{H0} = 100$ ng/ml for HMs, with an injection volume of $V_{\text{inj}} = 30$ ml. Two seepage velocities, $v = 0.087$ cm/s and $v = 0.260$ cm/s, are considered. The injection is performed from the top, and the initial injection velocity is maintained constant throughout the sand column. This setup accounts for the impact of consolidation deformation and stress paths induced by the external force on the porous media. The transport parameters used are listed in Table 7.1, based on references (Bai et al., 2021c). The physical sample consists of a saturated sand column with an initial void ratio of $e = 0.75$, a maximum stiffness $B_0 = 1 \times 10^6$ MPa, and a hardening parameter $\lambda_0 = 0.18$, representing a medium-dense soil with notable compressibility. This section focuses on the attachment of SPs and HMs at various depths and times along the sand column, while also examining the outflow concentration at the column's bottom.

Numerical calculations are performed using the commercial FEM software (COMSOL Multiphysics 6.0). The grid elements are calibrated for fluid dynamics, and a structured quadrilateral grid is chosen for the axial transport of SPs. The grid consists of 12 elements in the radial direction and 90 elements in the axial direction, totaling 1080 cells. Based on grid sensitivity analysis, a fine cell size is selected for high computational accuracy and low computational cost. A fully coupled computational method is employed to solve multi-field equations simultaneously, which ensures comprehensive solutions for the system's variables, despite higher convergence difficulty. This method also improves boundary condition handling and interaction modeling, reducing coupling errors between physical fields and enhancing the stability and convergence of the numerical solutions.

Mechanical deformation is solved through partial differential equations for mechanical deformation, while granular temperature conservation is addressed using ordinary differential equation method (DEM). Time discretization is achieved with the Euler backward differencing method, which is simple, stable, and effective for capturing dynamic behavior. The iterative process uses the automatic Newton method, adjusting the step size during the system solution, which is useful for nonlinear problems. A transient solver is used with a tolerance of 0.01, relying on residual reduction and time step selection for convergence. The initial time step of 0.001 is based on the sensitivity of the initial conditions. Adaptive time stepping uses an error estimation method to estimate the difference between results at the current time step and the exact solution. This allows the system to adjust the time step length to control error within a predefined range. The relative error in the computation is 0.001, indicating the error between the numerical solution and the theoretical value must be below 0.001. The error factor of 0.1 controls the adaptive time step and is associated with numerical and discretization errors, jointly ensuring the convergence and accuracy of the numerical solution.

TABLE 7.1 Calculation Parameters for Suspended Substance Transport

Mechanical and Thermal Parameters	Values	Transport Parameters	Values
Hardening parameter, λ_0	0.18	Diffusion coefficient of SPs, $D_P(m^2 \cdot s^{-1})$	2×10^{-4}
Maximum stiffness, B_0 (MPa)	1×10^6	Diffusion coefficient of HMs, D_H ($m^2 \cdot s^{-1}$)	3×10^{-4}
Parameter related to frictional property, χ	0.2	Viscosity coefficient of HMs, $\mu_H\left(Pa \cdot s\right)$	2.98×10^{-3}
Material parameter, c_1	0.5	Viscosity coefficient of SPs, $\mu_P\left(Pa \cdot s\right)$	4×10^{-3}
Material parameter, c_2 (s^2)	180	Attachment coefficient of solid matrix, $k_{Sd}(m^3 \cdot mol^{-1})$	1.38×10^{-5}
Material parameter, c_3 (s^2)	9×10^3	Detachment coefficient of solid matrix, $k_{Sr}(m^3 \cdot mol^{-1})$	9.84×10^{-6}
Material parameter, c_4 (kg/($m^3 \cdot s$))	100	Attachment coefficient of SPs, $k_{Pd}(m^3 \cdot mol^{-1})$	8.34×10^{-5}
Material parameter, c_5 (kg/($m^3 \cdot s$))	800	Detachment coefficient of SPs, $k_{Pr}(m^3 \cdot mol^{-1})$	6.89×10^{-6}
Specific heat capacity of solid matrix, c_S (J/(kg·K))	8.4×10^2	Transient equilibrium constant of attachment, $\kappa_d(m^3 \cdot mol^{-1})$ for $D_{50} = 5.1, 24.7\mu m$	6.8×10^5, 7.2×10^5
Specific heat capacity of HMs, c_H (J/(kg·K))	4.2×10^3	Transient equilibrium constant of detachment, κ_r ($m^3 \cdot mol^{-1}$) for $D_{50} = 5.1, 24.7\mu m$	4.8×10^{-5}, 4.3×10^{-5}
Hydraulic conductivity, k_W (m/s)	1×10^{-6}	Decay factor of detachment process, β_r	1.9
Specific heat capacity of SPs, c_P (J/(kg·K))	9.5×10^2	Decay factor of attachment process, β_d	0.05
		Reaction rate constant, λ (s^{-1})	2.2×10^{-2}

7.2.3.2 Impact of External Force on Seepage Transport

Under radial loading, the radial stress acting on any cross section of the soil column remains uniform along the radius. Therefore, the radial stress predicted by the CMF model matches the applied lateral pressure (i.e., $\sigma_r = p_a$; Figure 7.2a), which is consistent with the classical elastic solution and provides partial validation of the model's reliability. From Figure 7.3, the external force causes a significant reduction in the void ratio of the sand column. Starting from an initial value of 0.715, increasing the applied load (i.e., 100 kPa to 200 kPa then to 500 kPa) results in progressive compression of the soil skeleton, leading to corresponding decreases in void ratio *e* (0.657 to 0.614 then to 0.492). In essence, the CMF model successfully captures the rate-dependent nature of the loading process, reflecting the stress-path-dependent behavior of porous media. At relatively small loading levels (e.g., $\sigma_r = 100\,\text{kPa}$), the void ratio *e* exhibits an approximately linear relationship with $\log\left(\sigma_r\right)$. However, at higher loading magnitudes (e.g., $\sigma_r = 500\,\text{kPa}$), this relationship shows a distinct deviation after a certain stress level (around 30 kPa), beyond which it remains a more clearly linear trend. These trends indicate that soil consolidation deformation is not solely governed by elastoplastic responses described by continuum mechanics, but is fundamentally driven by microstructural evolution within the solid matrix, namely, particle rearrangement under mechanical loading. Figure 7.3 illustrates that the soil displays noticeable

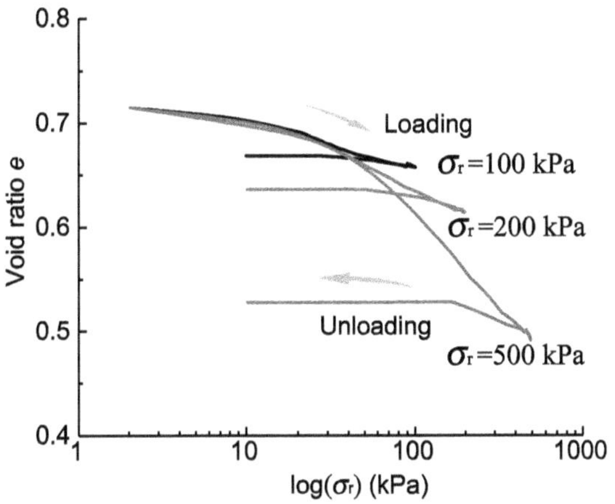

FIGURE 7.3 Consolidation deformation due to lateral loading–unloading.

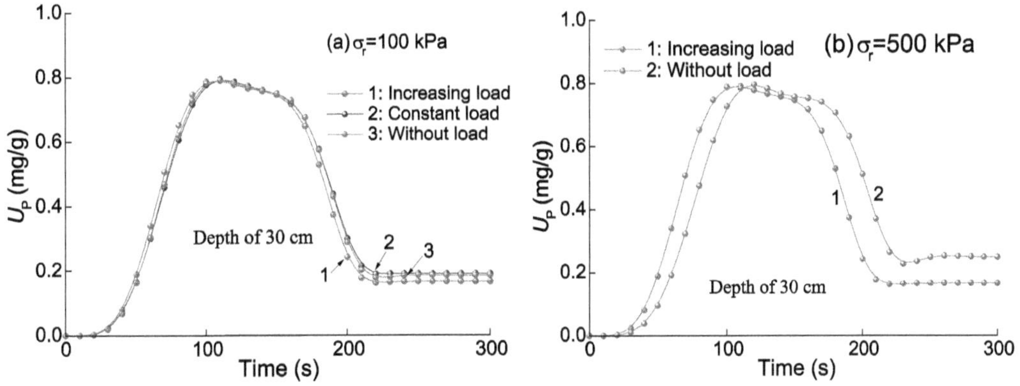

FIGURE 7.4 Attachment process of SPs with consolidation deformation. (a) $\sigma_r = 100\,\text{kPa}$. (b) $\sigma_r = 500\,\text{kPa}$.

rebound during unloading. The rebound effect is more pronounced at the beginning of unloading, particularly for $\sigma_r = 500\,\text{kPa}$, and transitions into a nearly linear trend as unloading progresses, highlighting the partial recoverability of the soil microstructure under decreasing external force.

Changes in the void ratio of the sand column directly impact the substance transport process. Figure 7.4 shows the temporal evolution processes of SPs concentration (U_P) under different lateral pressures (e.g., $\sigma_r = 100, 500$ kPa) and a constant seepage velocity ($v = 0.260$ cm/s) at a depth of 30 cm. Given that the injection time of the mixed solution ($t_{\text{inj}} = 120$ s) is shorter than the simulation time of the transport process ($t = 300$ s), the attachment of SPs at the same depth initially increases, then decreases to a stable value (i.e., residual concentration) over time. During the continuous injection process ($t = 0 - 120$ s), the attachment of SPs at the top of the column is greater than at the bottom. After the injection ends ($t > 120$ s), the residual concentration at the bottom becomes higher

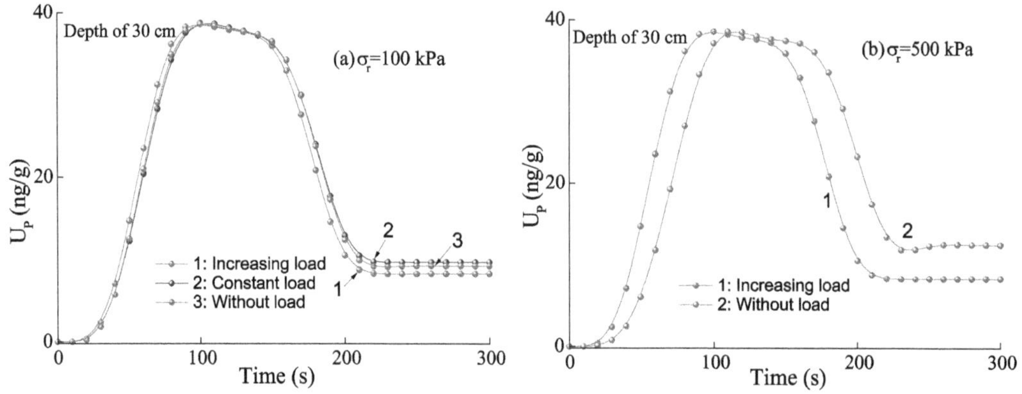

FIGURE 7.5 Attachment process of HMs with consolidation deformation. (a) $\sigma_r = 100$ kPa. (b) $\sigma_r = 500$ kPa.

than at the top and gradually stabilizes with time. When lateral pressure is increased from $\sigma_r = 100$ kPa to $\sigma_r = 500$ kPa (Figure 7.4), the attachment of SPs in the solid matrix increases. This is due to the higher lateral pressure causing pore compression in the soil column, which traps more SPs in the pores. Furthermore, the stabilization value for the case with gradually increasing load over time (path 1) is smaller than that for an instantaneous constant load. In this case, the void ratio of the sand column adjusts more quickly and reaches equilibrium sooner, leading to a higher residual concentration compared to the scenario with gradually increasing external force (Figure 7.4a).

From Figure 7.5, when there is no lateral displacement (i.e., constant void ratio), the attachment of HMs is greater than in the case where lateral pressure is applied during the continuous injection period ($t = 0-120$ s). This indicates that lateral pressure significantly affects HMs transport by reducing the void ratio. Over time, the concentration of HMs changes slightly, with the final residual concentration being higher than in the case without any applied load.

Figure 7.6 shows the breakthrough curves (BTCs) of SPs, represented by the outflow concentration C_{out} (i.e., effluent concentration; see Chapter 3) at a depth of 30 cm, under various lateral pressures and loading paths (no load, constant load, increasing load, and loading–unloading) at two different seepage velocities ($v = 0.087, 0.260$ cm/s).

Figure 7.6 illustrates that as lateral pressure increases (e.g., from $\sigma_r = 100$ to 500 kPa), the void ratio undergoes significant changes, amplifying the effect on the penetration process. From Figure 7.6b, the void ratio reaches a stable value more quickly under a constant load applied instantaneously (path 2) than under a gradually increasing load (path 1). Additionally, the outflow concentration is lower in the former case, and the detachment phenomenon is more noticeable, resulting in a broader effluent concentration gradient. With a higher seepage velocity of $v = 0.260$ cm/s, the peak value of the BTCs is higher. However, the smaller void ratio leads to a delay in reaching the peak value. Compared to the increasing load scenario, the unloading process following loading shows an increased peak value and an accelerated penetration process, with the peak occurring earlier in time due to the rising void ratio.

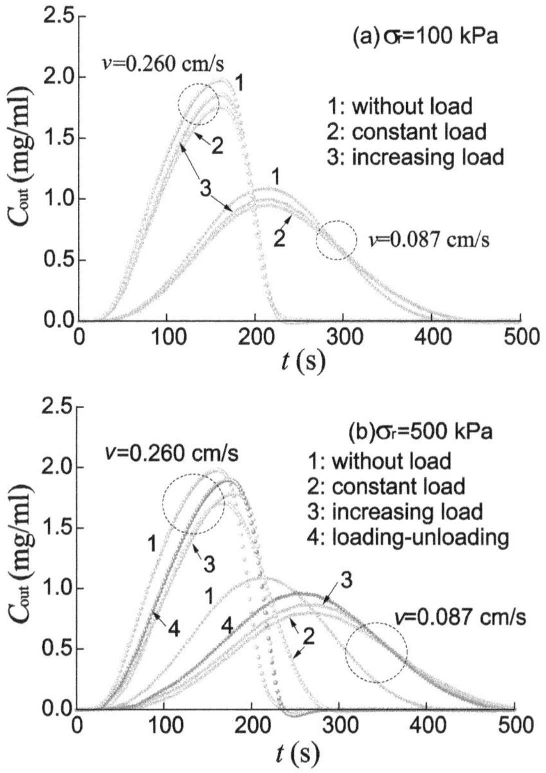

FIGURE 7.6 BTCs of SPs under varying radial forces. (a) $\sigma_r = 100$ kPa. (b) $\sigma_r = 500$ kPa.

Figure 7.7 shows the BTCs of HMs with time, exhibiting a similar trend to that of SPs. Notably, a more pronounced tailing effect is observed in the outflow concentration (C_{out}), indicating continued penetration of SPs even after the initial rising stage. As the seepage velocity increases, for both SPs and HMs (Figures 7.6 and 7.7), the peak of the BTCs rises significantly, while the trailing effect diminishes substantially. The presence of the tailing phenomenon is closely linked to the relative values of the detachment and attachment coefficients. Specifically, an increase in the detachment coefficient leads to a more noticeable tailing effect.

7.2.3.3 Temperature-Driven Effect on Seepage Transport

When a temperature difference exists, suspended substances migrate from the higher temperature region to the lower temperature region. Figures 7.8 and 7.9 show the evolution of SPs and HMs attachment over time under combined seepage flow ($v = 0.260$ cm/s) and temperature. A higher temperature ($T_1 = 80$ °C) is applied at the top of the sand column, while the bottom is maintained at ambient temperature ($T_0 = 20$ °C).

Temperature drives the transport of SPs through Brownian motion and facilitates the movement of HMs through thermal seepage and diffusion effects. This leads to a substantial decrease in the concentration of SPs and HMs remaining in the sand column. As lateral pressure increases, the void ratio decreases, which in turn raises the residual

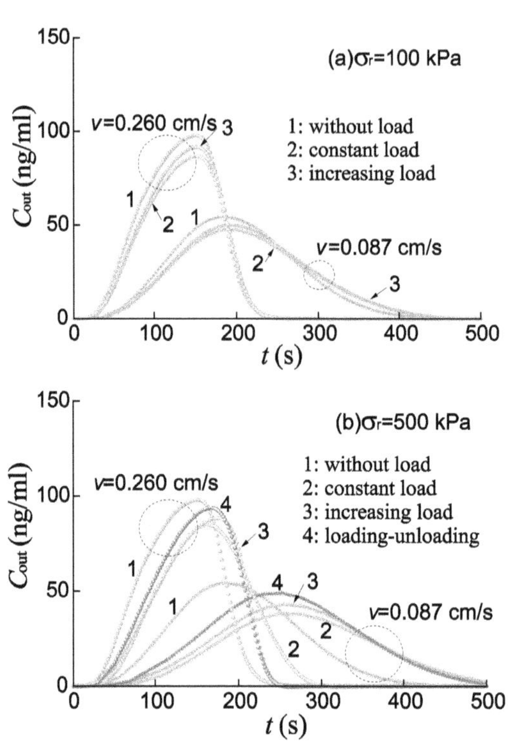

FIGURE 7.7 BTCs of HMs under varying radial forces. (a) $\sigma_r = 100$ kPa. (b) $\sigma_r = 500$ kPa.

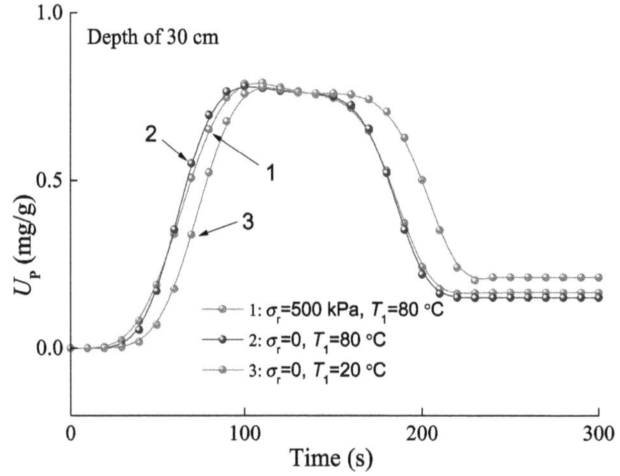

FIGURE 7.8 Temperature-driven BTCs of SPs: $\sigma_r = 0, 500$ kPa, $T_1 = 20, 80$ °C.

concentration (Figures 7.8 and 7.9). Furthermore, in the absence of temperature-driven effect (i.e., $T_1 = T_0 = 20$ °C), the residual concentrations remain the lowest during the continuous injection period ($t = 0-120$ s).

Figures 7.10 and 7.11 present the BTCs over time. As the temperature increases ($T_1 = 20, 60, 80$ °C), the peak concentration of SPs rises, and the time to reach the peak value

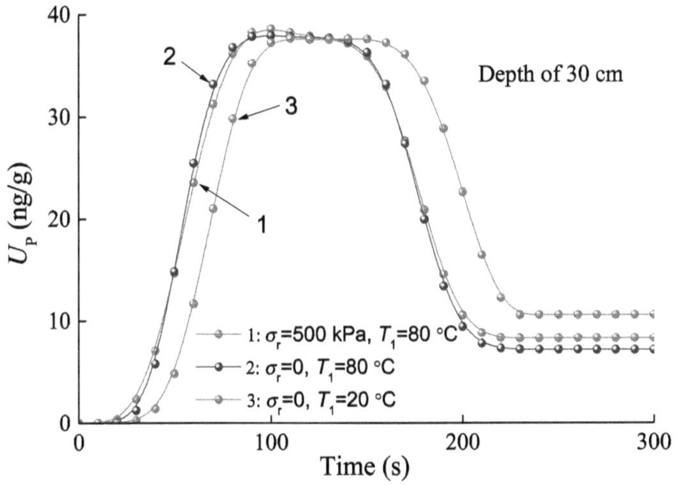

FIGURE 7.9 Temperature-driven BTCs of HMs: $\sigma_r = 0, 500$ kPa, $T_1 = 20, 80$ °C.

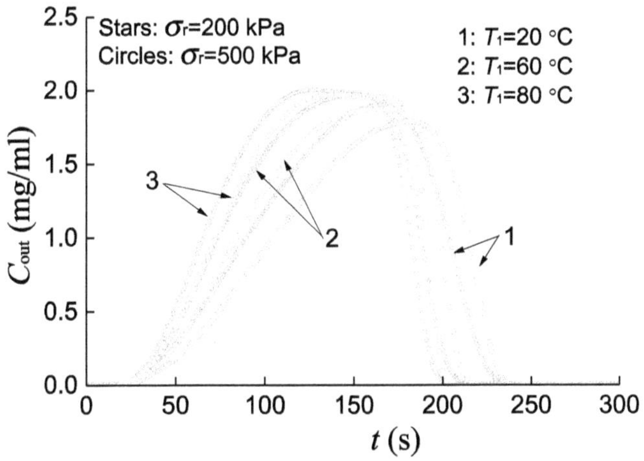

FIGURE 7.10 Temperature-driven BTCs of SPs at different temperatures.

shortens. This indicates that temperature driving accelerates the movement of suspended substances within the solid matrix, resulting in higher penetration peaks. At lower loads (e.g., $\sigma_r = 200$ kPa; Figure 7.11), the relatively larger void ratio causes the peak concentration of HMs to be significantly higher compared to higher loads (e.g., $\sigma_r = 500$ kPa), and the time to reach the peak is shorter.

7.2.4 Experimental Verification

The CMF model is validated by comparing its results with experimental data on seepage velocity and injection concentration over time, as reported in the literature (Bai et al., 2021d). Coupled multi-substance transport experiments are conducted using a one-dimensional soil column, with the specimen size shown in Figure 7.2a. The solid matrix

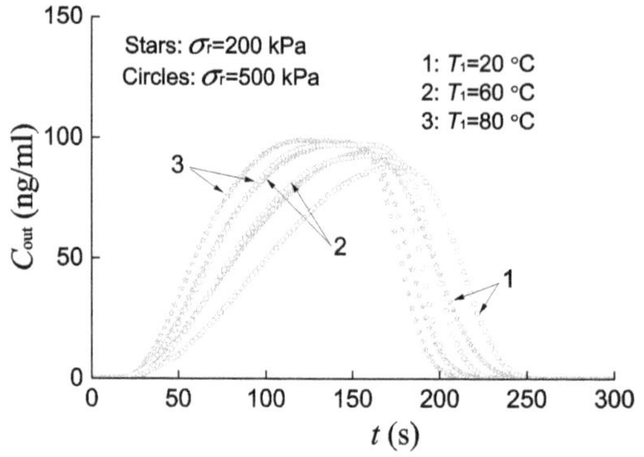

FIGURE 7.11 Temperature-driven BTCs of HMs at different temperatures.

consists of quartz sand with particle sizes ranging from 1 mm to 2 mm. The density and porosity of the soil column are provided in Section 4.3. The SPs are silicon powders with a median particle size of $D_{50} = 5.1\,\mu\text{m}, 24.7\,\mu\text{m}$ and their density matches that of Section 7.2. The HMs are lead ions (Pb^{2+}) with a solution of pH = 5.5. The parameters used for model verification are listed in Table 7.1.

Figure 7.12 displays the BTCs of SPs based on the flowing water pore volume (C_{out} is effluent concentration; V_P is pore volume; see Chapter 3) during changes in seepage velocity, with injected SPs concentrations ($C_{P0} = 0.5\,\text{mg/ml}, 2\,\text{mg/ml}$) and HMs concentrations ($C_{H0} = 100\text{ng/ml}, 200\,\text{ng/ml}$). Here, C_{P0} denotes the concentration of injected SPs, and C_{H0} refers to the concentration of injected HMs. Figures 7.12 and 7.13 show that the CMF model effectively captures the coupled transport of SPs and Pb^{2+} as seepage velocity changes. The model reveals an accelerated transport process (i.e., release effect) with increasing seepage velocity and a stronger attachment effect as the seepage velocity decreases. This indicates that Pb^{2+} enhances the attachment of SPs and lowers the concentration in BTCs, mainly due to the changes in the SPs surface's dielectric properties (Figure 7.12). Additionally, SPs facilitate the adsorption of Pb^{2+}, leading to a significant increase in penetration concentration (Figure 7.13). Both theoretical and experimental results indicate that, in the absence of SPs, the transport of Pb^{2+} is nearly unaffected by changes in water flow velocity, with HMs being fully dissolved in water, resulting in the lowest concentration observed.

Figure 7.14 presents the BTCs of Pb^{2+} as its injection concentration varies, while the concentrations of SPs are fixed at $C_{P0} = 0.5\,\text{mg/ml}, 2\,\text{mg/ml}$ and $D_{50} = 5.1\,\mu\text{m}, 24.7\,\mu\text{m}$ Figure 7.14 demonstrates that the CMF model effectively captures the coupled transport of SPs and HMs when $C_{H0} = 100\,\text{ng/ml}, 200\,\text{ng/ml}$. Specifically, the transport of HMs is significantly accelerated, as evidenced by their earlier appearance in the effluent. This effect becomes more noticeable as the SPs concentration decreases (e.g., $D_{50} = 5.1\,\mu\text{m}$). When the SPs concentration is increased further ($C_{P0} = 0.5\,\text{mg/ml}$ to $2\,\text{mg/ml}$, the comparison between Figure 7.14a and b), the presence of SPs enhances the transport of HMs even more.

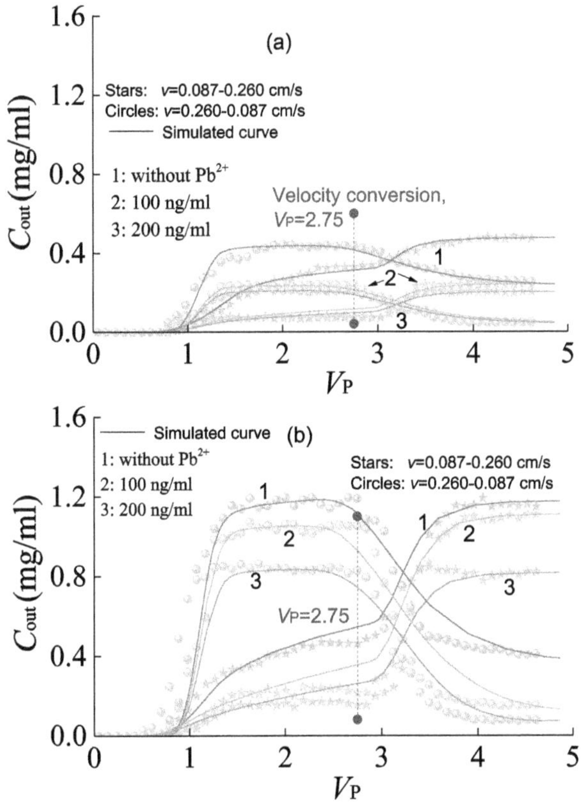

FIGURE 7.12 BTCs of SPs under varying seepage velocities. (a) $D_{50} = 5.1\,\mu\text{m}$, $C_{P0} = 0.5\,\text{mg/ml}$. (b) $D_{50} = 5.1\,\mu\text{m}$, $C_{P0} = 2\ \text{mg/ml}$.

7.3 HMs TRANSPORT IN UNSATURATED SOILS

7.3.1 Theory Framework

7.3.1.1 Mass, Momentum, and Energy Conservation

For coupled multi-substance flow (CMF) in unsaturated soils driven by seepage and temperature (Figure 7.15), the mass conservation equation is given by (Bai et al., 2025b)

$$\frac{\partial}{\partial t}\rho^{\alpha} - \nabla^2\left(D_{\alpha}\rho^{\alpha}\right) + \nabla_i\left(\rho^{\alpha}v^{\alpha S}\right) = \hat{\rho}^{\alpha} \tag{7.46}$$

where α represents the phase (S,W,G,H); $\hat{\rho}^{\alpha}$ and $v^{\alpha S}$ denote the mass conversion rate and transport velocity (i.e., attachment–detachment or liquid water evaporation process) between phases, respectively; D_{α} refers to the diffusion coefficient, relevant only for $\alpha = W,G,H$.

The equation for total momentum conservation is expressed as

$$\Sigma_{\alpha=S,W,G,H}\frac{\partial}{\partial t}(\rho^{\alpha}v^{\alpha S}) + \Sigma_{\alpha=S,W,G,H}\nabla_i(\sigma^{\alpha}) - \hat{p}^{\alpha} - \hat{\rho}^{\alpha}v^{\alpha S} = \Sigma_{\alpha=S,P,G,H}\rho^{\alpha}g \tag{7.47}$$

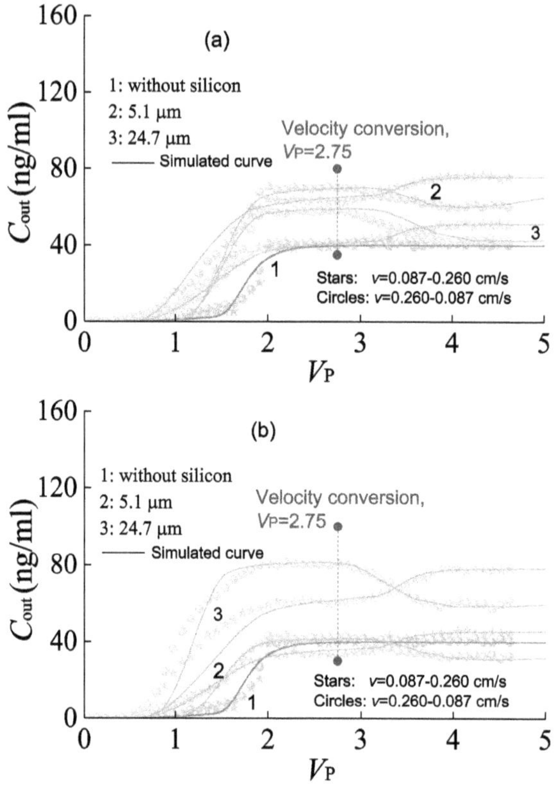

FIGURE 7.13 BTCs of HMs under varying seepage velocities. (a) $C_{P0} = 0.5\,\text{mg/ml}$, $C_{H0} = 0.5\,\text{ng/ml}$. (b) $C_{P0} = 2\,\text{mg/ml}$, $C_{H0} = 0.5\,\text{ng/ml}$.

where $\hat{p}^{\alpha}$ denotes the momentum exchange term and $\sum_{\alpha=S,L,G,H}(\hat{p}^{\alpha}) = 0$; σ^{α} represents the Cauchy stress tensor; g refers to the acceleration of gravity; $\hat{\rho}^{\alpha} v^{\alpha S}$ represents the material exchange on momentum.

The equation for total energy conservation is expressed as

$$\sum_{\alpha=S,W,G,H} \rho^{\alpha} \frac{\partial}{\partial t} e^{\alpha} - \hat{e}_{\alpha} + \hat{p}^{\alpha} \cdot v^{\alpha S} = \sum_{\alpha=S,W,G,H} n^{\alpha} \sigma_{kj}^{\alpha} \psi_{jk}^{\alpha} + \nabla_k \left(w v_k + \sigma_{kj} v_j - Q_k \right) - \sum_{\alpha=W,G,H} \rho^{\alpha} v_k^{\alpha S} \nabla_k e^{\alpha} \tag{7.48}$$

where Q_k represents the energy flow; e^{α} denotes the specific energy; ψ_{jk}^{α} refers to the deformation rate, given by $\psi_{jk}^{\alpha} = -\left(\nabla_j v_k^{\alpha} + \nabla_k v_j^{\alpha}\right)/2$; $\hat{e}_{\alpha}$ is the energy contribution rate from other phases; The velocities v_k^{WS}, v_k^{GS} and v_k^{HS} represent the relative flowing velocities of liquid water, gas and HMs with respect to the solid matrix.

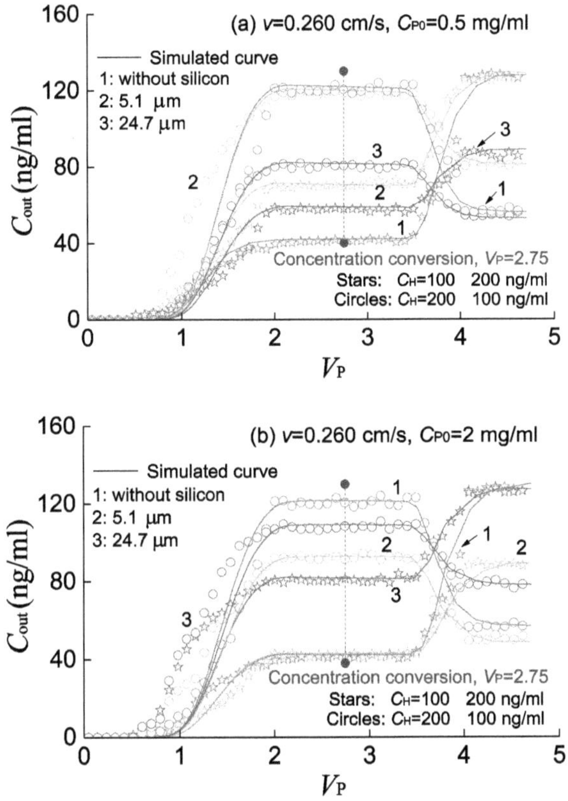

FIGURE 7.14 BTCs of HMs with varying injection concentrations. (a) v = 0.260 cm/s, C_{P0} = 0.5 mg/ml. (b) v = 0.260 cm/s, C_{P0} = 2 mg/ml.

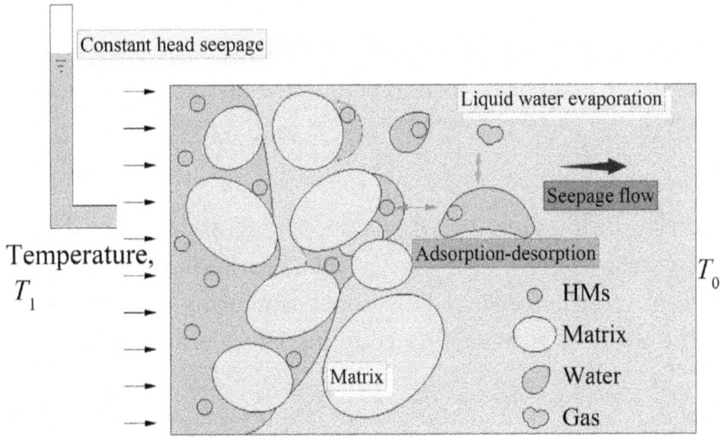

FIGURE 7.15 Seepage-temperature-driven multi-substance transport in unsaturated soil.

7.3.1.2 Thermodynamic Equation

For the entire system, the equation of entropy increase can be written as follows;

$$\Sigma_{\alpha=S,W,G,H}\rho^{\alpha}\frac{\partial}{\partial t}(\vartheta_{\alpha})=\frac{R}{T}+\nabla_{k}f_{k}-\rho^{W}v^{WS}\nabla_{k}(\vartheta_{W})-\rho^{G}v^{GS}\nabla_{k}(\vartheta_{G})-\rho^{H}v^{HS}\nabla_{k}(\vartheta_{H}) \quad (7.49)$$

For granular entropy increase, the thermodynamic equation is

$$\rho^{S}d_{t}(\vartheta_{g})=\frac{R_{g}}{T_{g}}-I_{g} \quad 7.50)$$

where ϑ_{α} represents the entropy capacity (the entropy per unit mass); R and R_{g} denote the generation rates of thermal entropy and granular entropy; T denotes the absolute temperature (°C); T_{g} denotes the granular temperature; f_{k} indicates the entropy conversion due to heat conduction; ϑ_{g} denotes the granular entropy capacity; I_{g} indicates the decay rate of granular entropy.

In the case of a representative element, its set of independent state-variables is represented as $A=\{\rho^{\alpha},U_{i}^{\alpha},\varepsilon_{ij}^{e},S,S_{g}\}$, in which ε_{ij}^{e} refers to the elastic strain, and S and S_{g} denote the thermal entropy and granular entropy. The corresponding conjugate set is $B=\{\mu_{ij}^{\alpha},v_{i}^{\alpha},\pi_{ij},T,T_{g}\}$, with π_{ij} representing the elastic stress, μ_{ij}^{α} representing the chemical potential energy, and m_{i}^{α} referring to the kinetic energy.

The thermodynamic identity, when expressed in terms of specific energy, leads to (Bai et al., 2025b)

$$\begin{aligned}\rho^{\alpha}de^{\alpha} = \sum_{\alpha=S,W,P,H}&\left[\mu_{ij}^{\alpha}d\rho^{\alpha}+\frac{1}{2}\left(v^{\alpha S}\right)^{2}d\rho^{\alpha}+T\theta_{\alpha}d\left(\rho^{\alpha}\right)-e^{\alpha}d\rho^{\alpha}+T\rho^{\alpha}d\left(\theta_{\alpha}\right)\right]\\&+\pi_{ij}d\varepsilon_{ij}^{e}+T_{g}\rho^{s}d\left(\theta_{g}\right)+T_{g}\theta_{g}d\left(\rho^{S}\right)\end{aligned} \quad (7.51)$$

In a CMF system, the dissipation force set C and the dissipation flow set D are expressed as $C=\{\pi_{ij},T_{g},\nabla_{i}T,\psi_{ij}^{S},\psi_{ij}^{W},\psi_{ij}^{G},\psi_{ij}^{H},v_{i}^{WS},v_{i}^{GS},v_{i}^{HS}\}$ and $D=\{Y_{ij},I_{g},q_{i},\sigma_{ij}^{S},\sigma_{ij}^{W},\sigma_{ij}^{G},\sigma_{ij}^{H},\chi_{i}^{WS},\chi_{i}^{PS},\chi_{i}^{HS}\}$, respectively.

The following relationships hold $Y_{ij}=\eta_{ijkl}\cdot\pi_{ij}$, $I_{g}=\gamma\cdot T_{g}$, and $q_{i}=\kappa_{ik}/T\nabla_{i}T+\theta_{ik}^{\alpha S}\cdot v_{i}^{\alpha S}$. The viscous stress of phase α is given by $\sigma_{ij}^{\alpha}=\lambda_{ijkl}^{\alpha}\cdot\psi_{ij}^{\alpha}$. Furthermore, $\chi_{i}^{\alpha S}=\beta_{ik}^{\alpha S}\cdot v_{i}^{\alpha s}+\theta_{ik}^{\alpha S}\nabla_{i}T$, where the first term indicates energy flow under isothermal conditions, and the second term describes the thermal permeability coupling effect for phase α (with $\alpha=W,P,H$). The terms $\eta_{ijkl},\gamma,\frac{\kappa_{ik}}{T},\lambda_{ijkl}^{S},\lambda_{ijkl}^{W},\lambda_{ijkl}^{G},\lambda_{ijkl}^{H},\beta_{ik}^{WS},\beta_{ik}^{GS}$ and β_{ik}^{HS} refer to the migration coefficients for the primary dissipative flows. The terms $\theta_{ik}^{WS},\theta_{ik}^{GS}$, and θ_{ik}^{HS} represent the coefficients for the coupling flow transport, known as the thermo-osmotic coefficients.

The rate of energy dissipation, denoted as $\Re_{W}$ is derived from the product of the dissipative force and flow:

$$\Re_W = Y_{ij}\pi_{ij} + I_g T_g + q_i \nabla_i T + \sigma_{ij}^S \psi_{ij}^S + \sigma_{ij}^W \psi_{ij}^W + \sigma_{ij}^G \psi_{ij}^G + \sigma_{ij}^H \psi_{ij}^H + \chi_i^{WS} v_i^{WS} + \chi_i^{GS} v_i^{GS} + \chi_i^{HS} v_i^{HS} \tag{7.52}$$

The generation of granular entropy $\Re_g$ can be expressed

$$\Re_g = \sigma_{ij}^g \psi_{ij}^g + T_g s \dot{S}_r + \sum_{\alpha=S,H} f\left[U_\alpha, d_t(C_\alpha)\right] \tag{7.53}$$

where σ_{ij}^g represents the dissipative force linked to the deformation ψ_{ij}^g at the *meso*-level, with $\psi_{ij}^g = \psi_{ij}^S = \frac{\partial}{\partial t}\varepsilon_{ij}$. Additionally, $\dot{S}_r$ represents the dissipative rate of the saturation degree S_r; s denotes the matrix suction. The energy dissipation due to the deviation of phase α (where $\alpha = S,H$) from equilibrium, caused by adsorption and desorption in the CMF system, is given by $f\left[U_\alpha, d_t(C_\alpha)\right] = T_g \Omega_\alpha d_t(C_\alpha)$. Here, $\Omega_\alpha = \xi_\alpha U_\alpha$ represents the dissipative flow related to multi-substance concentration, where ξ_α denotes the dissipation coefficient associated with adsorption and desorption, and U_α refers to the adsorption concentration.

By incorporating the mass, momentum, and energy conservation equations, along with the thermal entropy and granular entropy increase equations, into the thermodynamic identity (Eq. (7.51)), the following equation can be derived (Bai et al., 2025a):

$$\chi_i^{\alpha S} v^{\alpha S} = v_k^{\alpha S}\left(\rho^\alpha \mu^\alpha \nabla_i \rho^\alpha - \rho^\alpha \nabla_k e_\alpha - T\rho^\alpha \nabla_k \vartheta_\alpha\right) \quad (\alpha = W,G,H) \tag{7.54}$$

$$-\Sigma_{\alpha=S,W,G,H}\left[\rho^\alpha \mu^\alpha\left(\nabla_i^2 D_\alpha \rho^\alpha + \hat{\rho}^\alpha\right)\right] = \Sigma_{\alpha=S,W,G,H} f\left(U_\alpha, d_t C_\alpha\right) + T_g s \dot{S}_r \quad (\alpha = W,G,H) \tag{7.55}$$

where $\rho^\alpha \mu^\alpha \rho^\alpha = p^\alpha$ represents the pressure.

Given that there is no mass exchange between the water phase and HMs, the mass change rate of the water phase is zero, that is, $m^W = 0$. For solid matrix phase, the equation on the left side of Eq. (7.55) is $-\rho^S \mu^S m^S = f\left[U_{HS}, d_t\left(C_{HS}\right)\right]$, and for HMs phase, it becomes $-\rho^H \mu^H m^H = -f\left[U_{HS}, d_t\left(C_{HS}\right)\right]$. Thus, the mass change rate $\hat{\rho}^\alpha$ can be written as

$$\hat{\rho}^\alpha = m^\alpha = d_t\left(\rho_\alpha\right) = M_\alpha d_t\left(C_\alpha\right) \tag{7.56}$$

where $M_\alpha\left(\alpha = S,H\right)$ represents the molar mass.

The infiltration of pollutant ions in unsaturated soils is examined, where the transition between free-water and bound-water is neglected. When considering the flow of liquid water and the transport of water vapor, the mass exchange term between moisture and water vapor can be expressed using the phase-change relationship and Eq. (7.55):

$$\hat{\rho}^W = -\left(\rho^W \mu^W - \rho^G \mu^G\right)\hat{\rho}^{WG} \tag{7.57}$$

$$\hat{\rho}^G = \left(\rho^W \mu^W - \rho^G \mu^G\right)\hat{\rho}^{WG} \tag{7.58}$$

$$\hat{\rho}^S + \hat{\rho}^W + \hat{\rho}^G = T_g s \dot{S}_r \tag{7.59}$$

7.3.1.3 Temperature Field Equation

The specific energies are given by $e^S = c_S T + \left[w_e\left(\rho_\alpha, \varepsilon_{ij}^e, T\right) + w\left(\rho_\alpha, T_g\right) / \rho^S \right]$, $e^W = c_W T$, $e^G = c_G T$, and $e^H = c_H T$ (Bai et al., 2025b). Substituting these into the left side of the energy conservation equation (Eq. (7.48)) leads to

$$\begin{aligned} &\sum_{\alpha=S,W,G,H} \rho^\alpha d_t e^\alpha = \rho^S\left(c_S d_t T + \frac{1}{\rho^S} d_t w_e + \frac{1}{\rho^S} d_t w \right) + \sum_{\alpha=W,G,H} \rho^\alpha c_\alpha d_t T \\ &= \Sigma_{\alpha=S,W,G,H}\left(\rho^S c_S d_t T\right) + d_t\left(w_e + w\right) \end{aligned} \tag{7.60}$$

The right-hand side of Eq. (7.48) can be derived using the thermodynamic identity (Eq. (7.51)) and the expressions for migration coefficients:

$$\begin{aligned} &\nabla_k\left(w v_k + \sigma_{kj} v_j - Q_k\right) = \nabla_k\left[\left(\kappa_{ik} \nabla_i T / T + \theta_{ik}^{WS} v^{WS} + \theta_{ik}^{GS} v^{GS} + \theta_{ik}^{HS} v^{HS}\right) \cdot T\right] \\ &= \nabla_k\left(\kappa_{ik} \nabla_i T\right) + \nabla_k\left[\left(\theta_{ik}^{WS} v^{WS} + \theta_{ik}^{GS} v^{GS} + \theta_{ik}^{HS} v^{HS}\right) \cdot T\right] \end{aligned} \tag{7.61}$$

By applying Eqs. (7.60), (7.61), and (7.48), the temperature field equation is obtained:

$$\begin{aligned} &\sum_{\alpha=S,W,G,H} \rho^\alpha c_\alpha d_t T + d_t\left(w_e + w\right) = \nabla_k\left(\kappa_{ik} \nabla_i T\right) + \sum_{\alpha=W,G,H} v^{\alpha S} \nabla_k T\left(\theta_{ik}^{\alpha S} + \rho^\alpha c_\alpha\right) \\ &+ \Sigma_{\alpha=W,G,H} v^{\alpha S} \nabla_k\left(w_e + w\right) + \Sigma_{\alpha=S,W,G,H} n^\alpha \sigma_{kj}^\alpha \psi_{jk}^\alpha \end{aligned} \tag{7.62}$$

In unsaturated soils, the migration coefficient κ_{ik}, which represents thermal conductivity, is dependent on dry density and saturation. It is given by the following expression (Bai et al., 2020b):

$$\kappa_{ik} = \kappa_0 + \xi_{\mathrm{d}} \rho_{\mathrm{d}} + \xi_r S_r \tag{7.63}$$

where κ_0 denotes the thermal conductivity at the initial temperature and saturation; ρ_{d} denotes the dry density; ξ_{d} and ξ_r denote fitting parameters.

7.3.1.4 Ion Transport Equation

By applying the identity for the internal energy dissipation rate in a closed system (Eq. (7.54)), the resulting equation is derived as

$$\rho^\alpha \mu^\alpha \nabla_i \rho^\alpha - \left(\nabla_k \sigma_{il}^\alpha + \rho^\alpha \nabla_k e^\alpha\right) + T \rho^\alpha \nabla_k \vartheta_\alpha = \chi_i^{\alpha S} \qquad (\alpha = W, G, H) \tag{7.64}$$

The relationship between dissipation force v_i^{WS}, v_i^{GS}, v_i^{HS} and their corresponding dissipation flow χ_i^{WS}, χ_i^{GS}, χ_i^{HS} reflects the effect of the temperature gradient:

$$\chi_i^{\alpha S} = \theta_{ik}^{\alpha S} \nabla_k T + \beta_{ik}^{\alpha S} v_i^{\alpha S} \qquad (\alpha = W, G, H) \tag{7.65}$$

The specific energy e^W, e^G, e^H are determined by the first law of thermodynamics:

$$\nabla_k e^\alpha = T\nabla_k \vartheta_\alpha + \mu^\alpha \nabla_k \rho^\alpha \qquad (\alpha = W, G, H) \tag{7.66}$$

By substituting Eqs. (7.65) and (7.66) into Eq. (7.64), the following result can be derived:

$$\theta_{ik}^{\alpha S} \nabla_k T + \beta_{ik}^{\alpha S} v_i^{\alpha S} = -\nabla_k \sigma_{il}^\alpha \qquad (\alpha = W, G, H) \tag{7.67}$$

The force and pore pressures can be expressed (Bai et al., 2023):

$$\sigma_{kj}^W = \rho^W \mu^W \rho^W \delta_{ij} = p^W \tag{7.68}$$

Substituting Eq. (7.68) into the right-hand side of Eq. (7.67):

$$\beta_{ik}^{\alpha S} v_i^{\alpha S} = -\nabla_i \left(p^\alpha\right) - \theta_{ik}^{\alpha S} \cdot \nabla_k T \qquad (\alpha = W, G, H) \tag{7.69}$$

In Eq. (7.69), the migration coefficients are expressed as follows: $\theta_{ik}^{\alpha S}$ represents a parameter related to specific heat capacity; $\theta_{ik}^{WS} = \rho^W c_W \delta_{ik}$, $\theta_{ik}^{GS} = \rho^G c_G \delta_{ik}$, and $\theta_{ik}^{HS} = \rho^H c_H \delta_{ik}$. The migration coefficients β_{ik}^{WS}, β_{ik}^{GS}, and β_{ik}^{HS} are associated with the permeability coefficient. It is assumed that $\beta_{ik}^{WS} = \frac{u_W}{K_W \cdot k_W} \delta_{ik}$, $\beta_{ik}^{GS} = \frac{u_G}{K_G \cdot k_G} \delta_{ik}$, and $\beta_{ik}^{HS} = \frac{u_H}{K_H \cdot k_H} \delta_{ik}$. As a result, Eq. (7.69) can be expressed as

$$v_i^{\alpha S} = -\frac{K_\alpha \cdot k_\alpha}{u_\alpha} \nabla_i \left(p^\alpha\right) - \frac{K_\alpha \cdot k_\alpha}{u_\alpha} \theta_{ik}^{\alpha S} \cdot \nabla_k T \qquad (\alpha = W, G, H) \tag{7.70}$$

where K_α represents the permeability coefficient (mD); k_α represents the relative permeability; u_α refers to the viscosity coefficients (with $u_W = u_H$).

The deformation rate is defined as $\psi_{ij}^\alpha = -\left(\nabla_i v_j^\alpha + \nabla_j v_i^\alpha\right)/2$, and the time derivative of a flowing phase is given by $d_t = \partial_t + v^S \nabla_i$. For the solid skeleton, the equation becomes $d_t \varepsilon_{ij} = \psi_{ij}^S - \varepsilon_{ik} \nabla_j v_k - \varepsilon_{jk} \nabla_i v_k$. For small deformations, one has $d_t \varepsilon_{ij} = \psi_{ij}^S$. For HMs phase, Eq. (7.46) simplifies as

$$\frac{\partial}{\partial t} C^H + \nabla_i C^H v^{HS} = -C^H d_t \psi_{kk}^S + \nabla^2 D_H \rho^H + m^H \qquad (\alpha = H) \tag{7.71}$$

The distribution field for CMF is obtained by substituting Eq. (7.70) into Eq. (7.71):

$$\frac{\partial}{\partial t} C^H + \nabla_i \left\{ \left(C_H\right) \left[-\frac{K_H \cdot k_H}{u_H} \nabla_i \left(p^H\right) - \frac{K_H \cdot k_H}{u_H} \cdot \rho^H c_H \nabla_k T \right] \right\} + C_H d_t \psi_{kk}^S = \nabla^2 D_H C_H + m^H \tag{7.72}$$

During the transport of water-soluble HMs (e.g., Pb^{2+}), CMF (Figure 7.15) are assumed to move at the same seepage velocity as water ($v^W = v^H$). Considering the mass exchange between the solid matrix and HMs, $m^S = m^{HS}$ and $m^H = -m^{HS}$, as outlined in Eq. (7.56). The isothermal linear adsorption model is applied to describe the adsorption and desorption processes between HMs and the solid matrix ($\alpha = S$). Therefore, the mass exchange phase for HMs is given by

$$m^H = k_d \dot{C}_H - k_r \dot{C}_H \tag{7.73}$$

where k_r denotes the detachment coefficient, k_d denotes the attachment coefficient of solid matrix to HMs.

7.3.1.5 Flow of Liquid Water

Liquid water movement adheres to the principle of mass conservation, which states that the difference between the mass of water entering and exiting the soil equals the change in the water mass within the soil. The equation for water movement is derived from Eq. (7.46):

$$\frac{\partial}{\partial t}\theta_W - \nabla^2 \left(D_W \theta_W\right) + \nabla_i \left(\theta_W v^{WS}\right) = \hat{\rho}^W \tag{7.74}$$

where θ_W denotes the flow rate of liquid water, and D_W denotes the liquid water diffusivity.

The diffusivity of liquid water is given by the following expression (Bai et al., 2020b):

$$D_W = p^W K_W / \left(\rho^W g\right) \tag{7.75}$$

In Eq. (7.75), $p^W = -s + \rho^W gh$. In Eq. (7.70), the liquid water permeability coefficient K_W incorporates the effects of temperature and phase-change. It is defined as $K_W = k_W \cdot K_S$, where K_S denotes the saturated permeability coefficient and k_W denotes the relative permeability, commonly expressed as $k_W = S_r^{\,4}$. The saturated Saint-Euclidean coefficient is a function of the intrinsic permeability K_{int}:

$$K_S = K_{\text{int}} \rho^W g / u_w \tag{7.76}$$

The intrinsic permeability in Eq. (7.76) is expressed as

$$\log_{10}\left(K_{\text{int}}\right) = k_0 - k_1 \rho_d \tag{7.77}$$

where k_0 and k_1 refer to fitting parameters.

The dynamic viscosity coefficient in Eq. (7.76) is given by

$$u_w = 0.666(T + 46)^{-1.55} \times 10^{-6} \tag{7.78}$$

Substituting Eqs. (7.58), (7.70), and (7.75) into Eq. (7.74) results in the water movement equation:

$$\frac{\partial}{\partial t}\theta_W - \nabla^2\left(\frac{\theta_W p^W K_W}{\rho^W g}\right) + \nabla_i\left[\theta_W\left(-\frac{K_W \cdot k_W}{u_W}\nabla_i\left(p^W\right) - \frac{K_W \cdot k_W}{u_W}\cdot\rho^W c_W \cdot \nabla_k T\right)\right] = -\left(\rho^W \mu^W - \rho^G \mu^G\right)\hat{\rho}^{WG} \tag{7.79}$$

7.3.1.6 Flow of Water Vapor

As temperature increases, water vapor is generated and mixes with the existing gas in the soil. With the evaporation of liquid water from the pores, the governing equation for vapor flow, based on Eq. (7.46), is expressed as

$$\frac{\partial}{\partial t}\theta_G - \nabla^2\left(D_G\theta_G\right) + \nabla_i\left(\theta_G v^{GS}\right) = \hat{\rho}^G \tag{7.80}$$

where θ_G denotes the flow rate of water vapor and D_G denotes the water vapor diffusion rate.

The diffusion rate of water vapor is given by (Bai et al., 2020b)

$$D_G = p^G K_G V_W / V_G \tag{7.81}$$

where V_W and V_G represent the molar volumes of liquid water and water vapor, respectively.

Typically, the molar volume of liquid water is $V_W = 1.8\times10^{-5}\,\mathrm{m^3/mol}$. The molar volume of water vapor can be determined using the gas state equation:

$$V_G = \frac{R\left(T + 273.2\right)}{p_{\mathrm{atm}}} \tag{7.82}$$

Where $R = 8.314\,\mathrm{J/\left(mol\cdot{}^\circ C\right)}$ is the universal gas constant. For a typical pore structure connected to the atmosphere, it is assumed that the water vapor pressure matches the standard atmospheric pressure, $p_{\mathrm{atm}} = 0.1013\,\mathrm{MPa}$.

The permeability coefficient K_G in Eq. (7.81) is expressed as

$$K_G = \frac{K_{G,\mathrm{int}}\left(1 - S_r\right)^3}{u_G} \tag{7.83}$$

where the dynamic viscosity of vapor is given by $u_G = 1.8\times10^{-11}\,\mathrm{MPa\cdot s}$, the vapor permeability $K_{G,\mathrm{int}}$ is expressed as $\log_{10}\left(K_{G,\mathrm{int}}\right) = -k_{G0} - k_{G1}\rho_d$, where k_{G0} and k_{G1} are fitting parameters.

The vapor pressure is given by $p^G = p_{sat}RH$, where p_{sat} represents the saturated vapor pressure and RH represents the relative humidity. This relationship leads to

$$p_{sat} = 0.611\exp\left(17.27\frac{T}{T+237.2}\right) \tag{7.84}$$

$$RH = \exp\left[\frac{sV_W}{R(T+273.2)}\times 10^6\right] \tag{7.85}$$

In the liquid water injection process, only the initial seepage velocities of water and ions are provided, while the vapor seepage velocity is considered negligible. By substituting Eqs. (7.59) and (7.81) into Eq. (7.80), the equation for vapor flow is derived:

$$\frac{\partial}{\partial t}\theta_G - \nabla^2\left(\theta_G p^G K_G V_W / V_G\right) + \nabla_i\left[\theta_G\left(-\frac{K_G \cdot k_G}{u_G}\nabla_i\left(p^G\right) - \frac{K_G \cdot k_G}{u_G}\cdot \rho^G c_G \nabla_k T\right)\right]$$
$$= \left(\rho^W \mu^W - \rho^G \mu^G\right)\hat{\rho}^{WG} \tag{7.86}$$

7.3.1.7 Liquid Water Vapor Transformation due to Phase-Change

The circulating temperature influences not only molecular thermal motion but also the matrix suction of the soil, which varies repeatedly due to the cyclic movement of liquid water. This results in the soil-water characteristic curve (SWCC) exhibiting significant hysteresis. To describe this phenomenon, first-order functional expressions for the drying curve (TDC) and the wetting curve (TWC) are proposed, as outlined by Bai and Yang (2019):

$$\theta^{TD} = \theta_0^{TD} + \left(\theta_1^{TD} - \theta_0^{TD}\right)S_e \tag{7.87}$$

$$\theta^{TW} = \theta_0^{TW} + \left(\theta_1^{TW} - \theta_0^{TW}\right)S_e \tag{7.88}$$

where θ_0^{TD} and θ_0^{TW} represent the lower-bound values of TDC and TWC curves, respectively; θ_1^{TD} and θ_1^{TW} correspond to the upper-bound values of TDC and TWC curves; S_e is used to fit the TDC and TWC curves, and it is defined as follows:

$$S_e = \frac{\theta_W - \theta_r}{\theta_s - \theta_r} \tag{7.89}$$

where θ_r and θ_s refer to the drying water and the saturated water content.

The phase-change terms, $\hat{\rho}^W$ and $\hat{\rho}^G$, in both the water movement and the water vapor flow equation are further defined:

$$\hat{\rho}^W = \frac{\partial}{\partial t}\left(\theta^{TW} - \theta^{TD}\right) \tag{7.90}$$

$$\hat{\rho}^G = \frac{\partial}{\partial t}\left(\theta^{TD} - \theta^{TW}\right) \tag{7.91}$$

7.3.2 Model Parameters and Experimental Verification

7.3.2.1 Test Description and Numerical Method

The developed model was validated by comparing it with experimental data from the literature (Bai et al., 2025c). For this purpose, a one-dimensional silty soil column was used in CMF transport experiments. The dimensions of the column were 60 cm in length and 14.8 cm in diameter (Figure 7.16).

Initially, the volumetric moisture content of the soil column was set to $\theta_W = 15.8\%$, which refers to the volume of liquid water relative to the total volume of the soil. The initial gas content was $\theta_G = 36.2\%$, which represents the volume of gas in relation to the total volume of the soil ($\theta_W + \theta_G = n^V$). The initial temperature of the soil column was set to $T_0 = 20\ °C$, with the possibility of applying a higher temperature at the top surface (e.g., $T = 60\ °C$). The dry density of the soil was 1.3 g/cm^3. At the top surface, a constant water-head injection ($h_w = 2$ m) with HMs solution was used. The experiments were carried out under five different conditions: deionized water, Pb^{2+} solution (20 and 100 μg/ml), Cu^{2+} solution (100 μg/ml), and Cd^{2+} solution (100 μg/ml). A vertically placed soil column was employed to maintain one-dimensional seepage conditions. It is important to note that the influence of gravity was considered when analyzing the mechanisms involved.

The verification parameters are presented in Table 7.2. Numerical calculations are performed using COMSOL Multiphysics 6.0, with a structured quadrilateral grid of 2162 cells (16 radial elements and 94 axial elements) optimized for fluid dynamics. A fully coupled computational method is employed to solve multiple field equations simultaneously, improving accuracy, managing boundary conditions, and minimizing coupling errors.

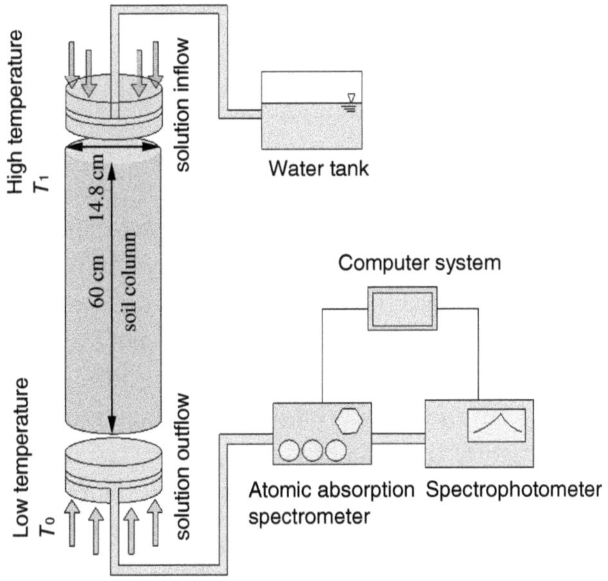

FIGURE 7.16 Physical model and boundary conditions.

TABLE 7.2 Calculation Parameters of CMF Transport

Mechanical and Thermal Parameters	Values	Transport Parameters	Values
Dry density, ρ_d (kg/m³)	1.3×10³	Fit parameters, ξ_d (W·m²/(kg·K))	0.6
Porosity, n^V	0.52	Fit parameters, ξ_r (W/(m·K))	1.1
Ion density, ρ_H (kg/m³)	1×10³	Diffusion coefficient of HMs, D_H (m²/s¹)	2.4×10^{-4}
Water density, ρ_W (kg/m³)	1×10³	Viscosity coefficient of HMs, u_H (Pa·s)	2.98×10^{-3}
Gas density, ρ_G (kg/m³)	1.2×10³	Adsorption coefficient of solid matrix, k_d (m³/mol)	6.52×10^{-5}
Specific heat capacity of solid matrix, c_S (J/(kg·K))	8.4×10^2	Desorption coefficient of solid matrix, k_r (m³/mol)	2.39×10^{-5}
Specific heat capacity of HMs, c_H (J/(kg·K))	4.2×10^3	Fit parameters, k_0	3.1
Specific heat capacity of water, c_W (J/(kg·K))	4.2×10^3	Fit parameters, k_1 (cm³/g)	2.17
Specific heat capacity of water, c_G (J/(kg·K))	1×10^3	Fit parameters, k_{G0}	1.51
Initial thermal conductivity, κ_0 (W/(m·K))	1	Fit parameters, k_{G1} (cm³/g)	6.05

Temperature conservation is handled through ordinary differential equations. The Euler backward differencing technique is applied to effectively capture dynamic behavior, while an automatic Newton method supports iterative solutions and adjusts step sizes for nonlinear problems. A transient solver is set with a tolerance of 0.01, and adaptive time stepping ensures that the relative error remains below 0.001. The adaptive time step is controlled by an error factor of 0.1, which helps guarantee the convergence and accuracy of the numerical solution.

7.3.2.2 Experimental Verification

Figure 7.17 displays the variation in moisture content over time at different depths (z = 37.5 cm and z = 52.56 cm). A comparison between the experimental data and simulation curves shows that the developed model matches the experimental results well. As the HMs solution is injected, the moisture content increases gradually from the initial value of 15.8% until the injection is stopped. Figure 7.17a depicts the moisture content changes at depths of 37.5 and 52.5 cm from the top of the column, which shows that points closer to the surface reach the moisture imbalance stage earlier, with the moisture increase beginning at t = 19 h for z = 37.5 cm and at t = 59 h for z = 52.5 cm. It is observed that the addition of HMs increases the moisture rise rate, even though the final moisture content is lower. This can be explained by the enhanced solute potential in the rehydrated organic matter, which reacts with mineral components, reducing the water retention ability of the soil particles. As a result, during the absorption process, water tends to flow from the region of lower matric potential to those with higher matric potential, meaning the presence of ions accelerates the moisture increase.

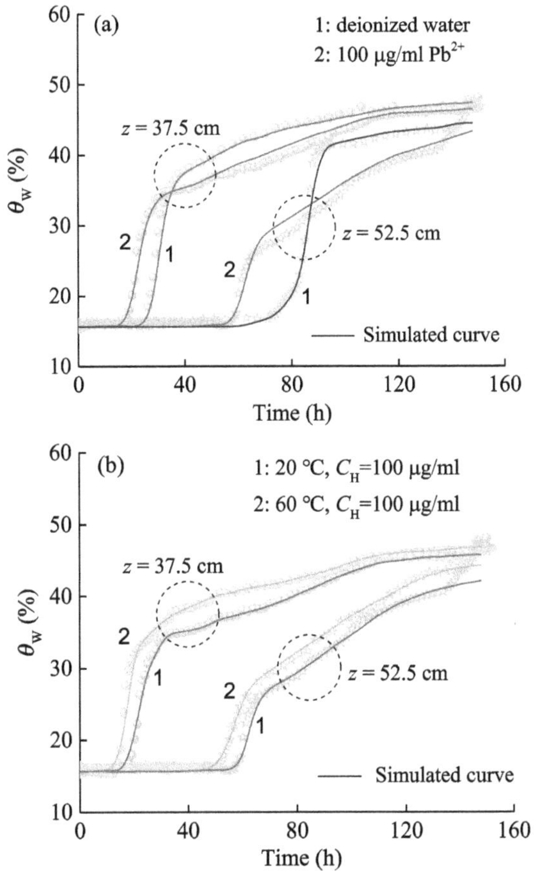

FIGURE 7.17 Moisture movement over time. (a) Different distances. (b) Different temperatures.

From Figure 7.17b, at a consistent depth and HMs concentration (z = 52.5 cm), higher temperatures at the soil column's top surface cause a quicker increase in moisture content. This indicates that as temperature rises (e.g., from 20 °C to 60 °C), the water molecules gain more internal energy. As a result, liquid water in the soil transitions more easily from bound-water to free-water, driven by a strong thermodynamic force. This leads to a significant acceleration in the rate of moisture movement.

Figure 7.18 illustrates the BTCs of HMs, showing a close match between the simulation results and experimental data. Initially, the HMs concentration is zero (Figure 7.18), leading to no concentration when transport starts. As the injected concentration increases, the outflow concentration also rises and eventually stabilizes. The curve shows a sharp increase, followed by a decrease, as it reaches stability. This behavior may be influenced by the solution injection rate. A high seepage velocity could cause the accumulation of HMs concentration. However, once the injection ceases and the porous medium's attachment-capacity is considered, the concentration of HMs decreases and eventually levels off.

From Figure 7.18a, it is observed that as the injected HMs concentration increases, the outflow concentration also rises. Changes in the attachment–detachment coefficients

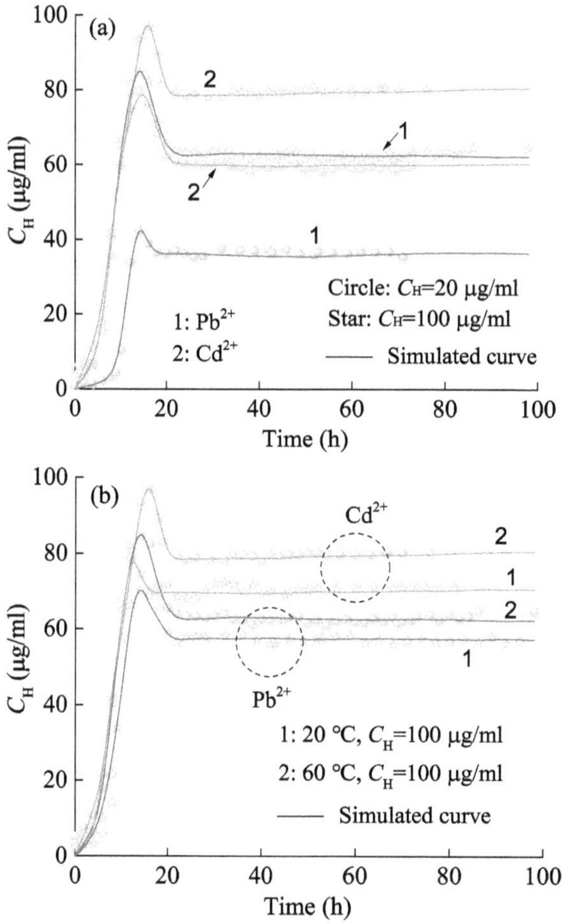

FIGURE 7.18 BTCs of HMs over time. (a) Different injected concentrations. (b) Different temperatures.

suggest that the medium shows a greater attachment-capacity for Pb^{2+} than for Cd^{2+}. In Figure 7.17b, the transport of HMs is shown to be influenced by thermal diffusion and permeation. At a constant depth and HMs concentration (z = 52.5 cm), higher temperatures at the top of the soil column speed up HMs transport. Additionally, at shallower depths (z = 37.5 cm < z = 52.5 cm), a higher steady-state HMs concentration is observed, emphasizing the significant impact of soil depth, porosity, and transport pathways on the process.

7.3.3 Multi-Substance Flow Transport under Seepage-Temperature

7.3.3.1 Calculation Model

The physical model shown in Figure 7.16 is used for further discussion. The seepage flow and temperature load are applied from the top surface (T_1) to the bottom surface (T_0). The concentration of injected HMs increases exponentially with time:

$$C_i(t) = C_0[1 - \exp(-\frac{\beta t}{M})] \tag{7.92}$$

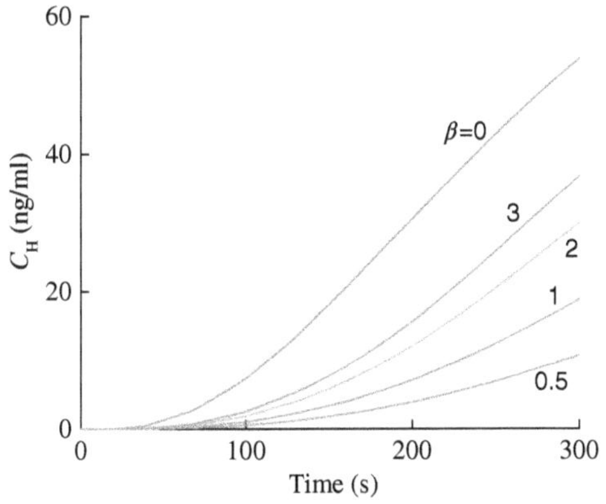

FIGURE 7.19 Sensitivity analysis of the degradation parameter.

where C_0 refers to the initial concentration, M refers to the time period, and β refers to the degradation parameter.

For simplicity, the dimensionless concentration C/C_0 is introduced. The initial HMs injection concentration is $C_0 = 100$ ng / ml, and the injection volume is $V_{\text{inj}} = 30$ ml. Two seepage velocities are discussed: $v = 0.087$ cm/s and 0.260 cm/s. The initial seepage velocity remains constant within the soil column. Transport parameters are detailed in Table 7.2. The moisture content of the soil column is θ_W = 52% when saturated.

Figure 7.19 shows that the BTCs of HMs at the bottom of the soil column increase with time but decrease as the degradation parameter β becomes larger. From Eq. (7.92), when $\beta = 0$, the injected HMs concentration remains constant (i.e., constant injection concentration, $C_i(t) = C_i$), resulting in the highest effluent concentration. As β increases (i.e., variable injection concentration, C_i), the effluent concentration still rises over time but remains lower than the case with $\beta = 0$ (Figure 7.19). In later simulations, the scenarios of constant injection concentration and variable injection concentration corresponding to Eq. (7.92) are evaluated at $\beta = 0$ and $\beta = 2$, respectively.

7.3.3.2 Injection Concentration Variation with Time

Figure 7.20 shows the moisture variations at various seepage velocities ($v_0^{HS} = v_0^{WS} =$ 0.087 and 0.260 cm/s) and varying HMs injection concentrations at depths of 5, 20, and 60 cm. The initial moisture content is set at 15.8%, with $T_1 = T_0 = 20\,^\circ\text{C}$. As the injected solution increases, the moisture content rises, and the moisture content at the top of the soil column approaches saturation over time (Figure 7.20a; reaching $\theta_W = 52\%$).

From Figure 7.20b, an increase in seepage velocity promotes a rise in moisture content. This suggests that higher seepage velocity enhances the hydrodynamic effect, allowing the injected solution to fill the soil more quickly. From Figure 7.20c, the injection of HMs significantly influences the moisture content variation. As HMs concentration increases (from variable to constant concentration), the adsorption of HMs by the solid matrix intensifies, resulting in a higher residual of HMs in the soil and an increase in moisture content.

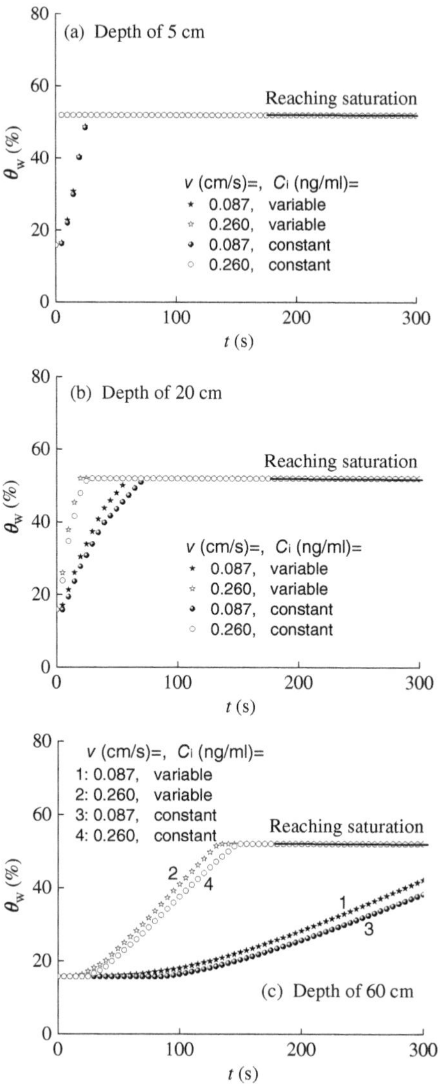

FIGURE 7.20 Moisture content variations. (a) z = 5 cm. (b) z = 20 cm. (c) z = 60 cm.

Figure 7.21 displays the BTCs of HMs for different seepage velocities ($v_0^{HS} = v_0^{WS} = 0.087$ and 0.260 cm/s) and varying injection concentrations of HMs at depths of 5, 20, and 60 cm. In saturated soils, seepage velocity has little impact on the BTCs of HMs. However, in unsaturated soils, increasing seepage velocity enhances hydrodynamic effects, which accelerates the rise in moisture content. This leads to a quicker reduction in matrix suction and higher moisture content in the effluent (Figure 7.20), causing an increase in HMs concentration.

7.3.3.3 Different Temperature Gradient

Figure 7.22 illustrates the moisture migration caused by temperature gradient ($\Delta T = T_1 - T_0 = 0, 30°C, 60°C, T_0 = 20\ °C$. As the HMs solution is injected from the top

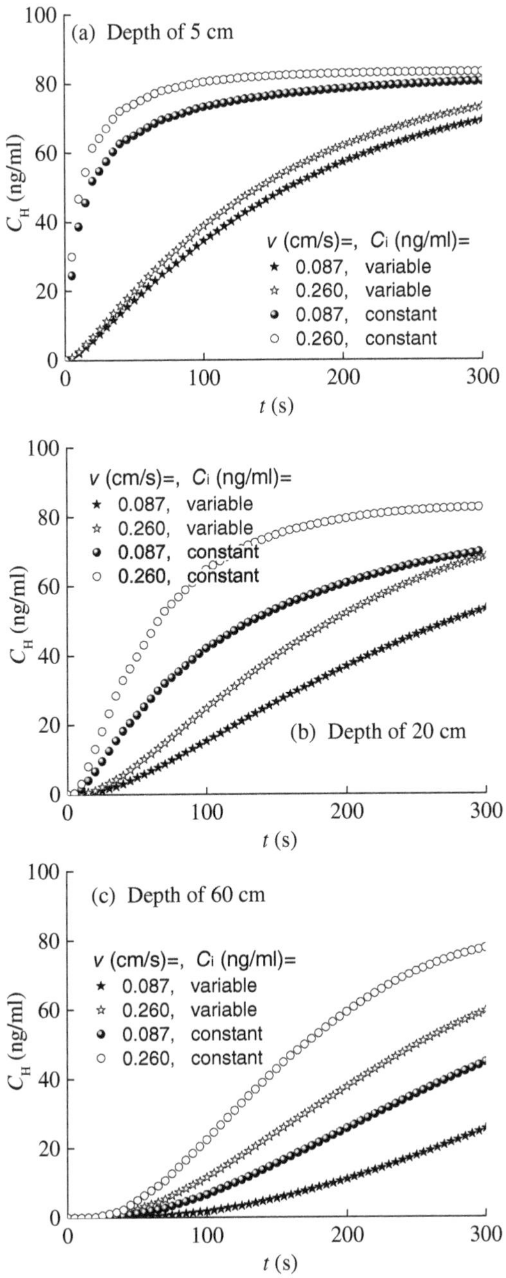

FIGURE 7.21 BTCs of HMs. (a) $z = 5$ cm. (b) $z = 20$ cm. (c) $z = 60$ cm.

surface, the soil column gradually approaches saturation (Figure 7.22a), with the soil nearer to the surface reaching saturation more rapidly. It is apparent that a faster injection of HMs solution results in quicker moisture transport and earlier saturation (Figure 7.22b). Additionally, increasing seepage velocity improves heat transfer to the bottom of the soil column (Figure 7.22c). This not only accelerates the saturation process but also enhances

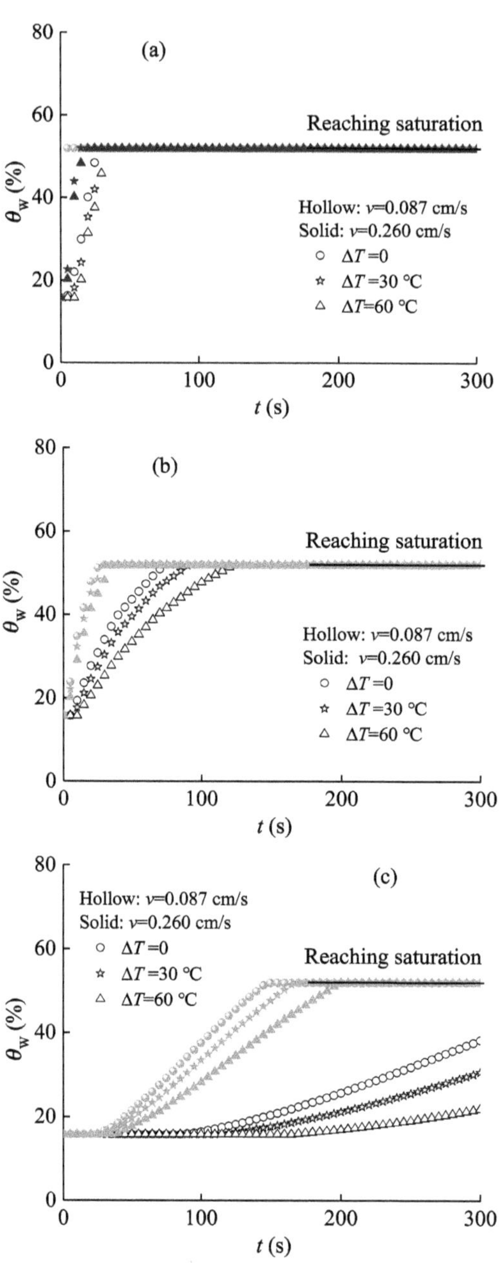

FIGURE 7.22 Moisture content evolution caused by temperature. (a) $z = 5$ cm. (b) $z = 20$ cm. (c) $z = 60$ cm.

the phase transition of moisture into water. From Figure 7.22, the higher temperature gradient accelerates the conversion of liquid water into vapor, reducing moisture and slowing the saturation process.

Figure 7.23 presents the gas evolution at various depths under different temperature gradients and seepage velocities. The injection of the HMs solution expels gas from the soil,

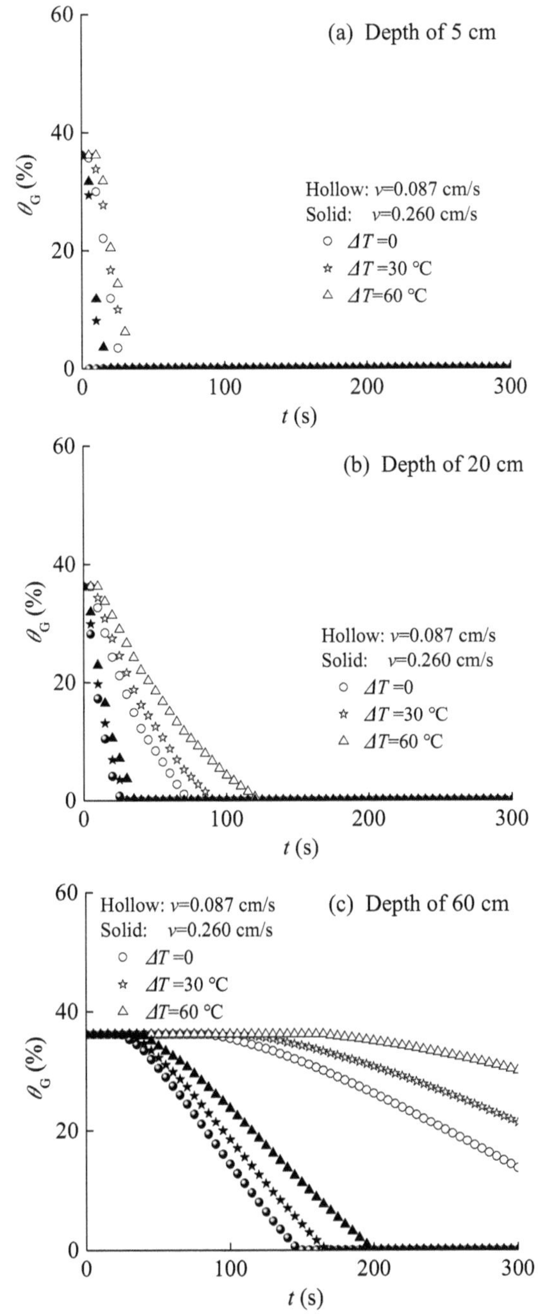

FIGURE 7.23 Volumetric air content evolution caused by temperature. (a) $z = 5$ cm. (b) $z = 20$ cm. (c) $z = 60$ cm.

with gas being released more quickly near the injection interface. Figure 7.23b illustrates that higher seepage velocities speed up gas expulsion, reducing θ_G toward zero. The high temperature shows that higher seepage velocities facilitate faster heat penetration into the soil (Figures 7.23c and 7.22c), accelerating the phase transition from moisture to gas.

References

Abbar B, Alem A, Pantet A, Ahfir Nasre-Dine, Marcotte S, Wang H, Duchemin B. Effect of natural geotextile on the cotransport of heavy metals (Cu^{2+}, Pb^{2+}, and Zn^{2+}) and kaolinite particles. *Environmental Technology*, 2019, 42(4): 558–570

Abramowitz M, Stegun I A. *Handbook of Mathematical Functions with Formulas, Graphs, and Mathematical Tables.* National Bureau of Standards Applied Mathematics Series 55. Tenth Printing. 1972

Ahfir N D, Hammadi A, Alem A, Wang H Q, Bras G, Ouahbi T. Porous media grain size distribution and hydrodynamic forces effects on transport and deposition of suspended particles. *Journal of Environmental Sciences*, 2017, 53: 161–172

Alem A, Ahfir N D, Elkawafi A, Wang Q. Hydraulic operating conditions and particle concentration effects on physical clogging of a porous medium. *Transport in Porous Media*, 2015, 106: 303–321

Alshaer A W, Rogers B D, Li L. Smoothed Particle Hydrodynamics (SPH) modelling of transient heat transfer in pulsed laser ablation of Al and associated free-surface problems. *Computational Materials Science*, 2017, 127: 161–179

Arab D, Pourafshary P, Ayatollahi S, Habibi A. Remediation of colloid-facilitated contaminant transport in saturated porous media treated by nanoparticles. *International Journal of Environmental Science and Technology*, 2014, 11: 207–216

Ataie-Ashtiani B, Nik-Khah A. Impulsive waves caused by subaerial landslides. *Environmental Fluid Mechanics*, 2008, 8: 263–280

Bai B. Consolidation solutions of a saturated porothermoelastic hollow cylinder with infinite length. *Engineering*, 2010, 2(1): 37–45

Bai B. Fluctuation responses of saturated porous media subjected to cyclic thermal loading. *Computers and Geotechnics*, 2006, 33(8): 396–403

Bai B, Bai F, Hou J. The migration process and temperature effect of aqueous solutions contaminated by heavy metal ions in unsaturated silty soils. *Heliyon*, 2024, 10: e30458

Bai B, Bai F, Li X, Nie Q, Jia X, Wu H. The remediation efficiency of heavy metal pollutants in water by industrial red mud particle waste. *Environmental Technology & Innovation*, 2022a, 28: 102944

Bai B, Jiang S, Liu L, Li X, Wu H. The transport of silica powders and lead ions under unsteady flow and variable injection concentrations. *Powder Technology*, 2021a, 387: 22–30

Bai B, Li H, Xu T, Chen X. Analytical solutions for contaminant transport in a semi-infinite porous medium using the source function method. *Computers and Geotechnics*, 2015, 69: 114–123

Bai B, Long F, Rao D Y, Xu T. The effect of temperature on the seepage transport of suspended particles in a porous medium. *Hydrological Processes*, 2017a, 31(2): 382–393

Bai B, Nie Q, Wu H, Hou J. The attachment-detachment mechanism of ionic/nanoscale/microscale substances on quartz sand in water. *Powder Technology*, 2021b, 394: 1158–1168

Bai B, Nie Q, Zhang Y, Wang X, Hu W. Cotransport of heavy metals and SiO_2 particles at different temperatures by seepage. *Journal of Hydrology*, 2021c, 597: 125771

Bai B, Rao D, Chang T, Guo Z. A nonlinear attachment-detachment model with adsorption hysteresis for suspension-colloidal transport in porous media. *Journal of Hydrology*, 2019a, 578: 124080

Bai B, Rao D Y, Xu T, Chen P P. SPH-FDM boundary for the analysis of thermal process in homogeneous media with a discontinuous interface. *International Journal of Heat and Mass Transfer*, 2018, 117: 517–526

Bai B, Ren Y, Rao D. The transport of solid particle suspension at different alkalinities in saturated porous media. *Journal of Porous Media*, 2020a, 23(3): 207–218

Bai B, Wang J, Zhai Z, Xu T. The penetration processes of red mud filtrate in a porous medium by seepage. *Transport in Porous Media*, 2017b, 117(2): 207–227

Bai B, Wang Y, Rao D, Bai F. The effective thermal conductivity of unsaturated porous media deduced by pore-scale SPH simulation. *Frontiers in Earth Science*, 2022b, 10: 943853

Bai B, Wu H, Nie Q, Liu J, Jia X. Granular thermodynamic migration model suitable for high-alkalinity red mud filtrates and test verification. *International Journal for Numerical and Analytical Methods in Geomechanics*, 2025a, 49: 1530–1543

Bai B, Wu H, Wu N, Zhang B. Granular thermodynamic ideology on the heavy metal migration in unsaturated soils driven by seepage-temperature. *Journal of Hydrology*, 2025b, 661: 133738

Bai B, Wu H, Zhou R, Wu N, Zhang B. A granular thermodynamic framework-based coupled multiphase-substance flow model considering temperature driving effect. *Journal of Rock Mechanics and Geotechnical Engineering*, 2025c, 17: 5816–5828

Bai B, Xu T, Guo Z. An experimental and theoretical study of the seepage migration of suspended particles with different sizes. *Hydrogeology Journal*, 2016, 24(8): 2063–2078

Bai B, Xu T, Nie Q, Li P. Temperature-driven migration of heavy metal Pb^{2+} along with moisture movement in unsaturated soils. *International Journal of Heat and Mass Transfer*, 2020b, 153: 119573

Bai B, Yang G. Thermo-hydro-mechanical model for unsaturated clay soils based on granular solid hydrodynamics theory. *International Journal of Geomechanics*, 2019b, 19: 04019115

Bai B, Zhai Z, Rao D. The seepage transport of heavy metal Pb^{2+} through sand column in the presence of silicon powders. *Journal of Hydrology and Hydromechanics*, 2019c, 67: 349–358

Bai B, Zhang J, Liu L, Ji Y. The deposition characteristics of coupled lead ions and suspended silicon powders along the migration distance in water seepage. *Transport in Porous Media*, 2020c, 134(3): 707–724

Bai B, Zhou R, Cai G, Hu W, Yang G. Coupled thermo-hydro-mechanical mechanism in view of the soil particle rearrangement of granular thermodynamics. *Computers and Geotechnics*, 2021d, 137: 104272

Bai B, Zhou R, Yang G, Zou W, Yuan W. The constitutive behavior and dissociation effect of hydrate-bearing sediment within a granular thermodynamic framework. *Ocean Engineering*, 2023, 268: 113408

Basha H A, Selvadurai P S A. Heat-induced moisture transport in the vicinity of a spherical heat source. *International Journal for Numerical and Analytical Methods in Geomechanics*, 1998, 22: 969–981

Bedrikovetsky P, You Z, Badalyan A, Osipov Y, Kuzmina L. Analytical model for straining-dominant large-retention depth filtration. *Chemical Engineering Journal*, 2017, 330: 1148–1159

Bedrikovetsky P, Zeinijahromi A, Siqueira F, Furtado C, Souza A L. Particle detachment under velocity alternation during suspension transport in porous media. *Transport in Porous Media*, 2012, 91: 173–197

Bekhit H M, EI-Kordy M A, Hassan A E. Contaminant transport in groundwater in the presence of colloids and bacteria: model development and verification. *Journal of Contaminant Hydrology*, 2009, 108: 152–167

Bennacer L, Ahfir N D, Alem A, Wang H Q. Coupled effects of ionic strength, particle size, and flow velocity on transport and deposition of suspended particles in saturated porous media. *Transport in Porous Media*, 2017, 118(3): 251–269

Bennacer L, Ahfir N D, Bouanani A, Alem A, Wang H Q. Suspended particles transport and deposition in saturated granular porous medium: particle size effects. *Transport in Porous Media*, 2013, 100(3): 377–392

Borazjani S, Bedrikovetsky P. Exact solutions for two-phase colloidal-suspension transport in porous media. *Applied Mathematical Modelling*, 2017, 44: 296–320

Bradford S A, Kim H, Shen C, Sasidharan S, Shang J. Contributions of nanoscale roughness to anomalous colloid retention and stability behavior. *Langmuir*, 2017, 33(38): 10094–10105

Bradford S A, Kim H N, Haznedaroglu B Z, Torkzaban S, Walker S L. Coupled factors influencing concentration-dependent colloid transport and retention in saturated porous media. *Environmental Science and Technology*, 2009, 43(18): 6996–7002

Chang C, Wang M, Chang T, Lin C, Chen Y. Transport modeling of copper and cadmium with linear and nonlinear retardation factors. *Chemosphere*, 2001, 43(8): 1133–1139

Chen J S, Chen J T, Liu C W, Liang C P, Lin C W. Analytical solutions to two-dimensional advection-dispersion equation in cylindrical coordinates infinite domain subject to first- and third-type inlet boundary conditions. *Journal of Hydrology*, 2011, 405(3–4): 522–531

Chequer L, Bedrikovetsky Pavel, Badalyan A, Gitis V. Water level and mobilisation of colloids in porous media. *Advances in Water Resources*, 2020, 143: 103670

Chequer L, Carageorgos T, Naby M, Hussaini M, Lee W, Bedrikovetsky P. Colloidal detachment from solid surfaces: phase diagrams to determine the detachment regime. *Chemical Engineering Science*, 2021, 229: 116146

Chrysikopoulos C V, Katzourakis V E. Colloid particle size-dependent dispersivity. *Water Resources Research*, 2015, 51: 4668–4683

Chrysikopoulos C V, Syngouna V I. Effect of gravity on colloid transport through water-saturated columns packed with glass beads: modeling and experiments. *Environmental Science and Technology*, 2014, 48: 6805–6813

Cleary P W, Monaghan J J. Conduction modelling using smoothed particle hydrodynamics. *Journal of Computational Physics*, 1999, 148: 227–264

Corapcioglu M Y, Jiang S. Colloid-facilitated groundwater contaminant transport. *Water Resources Research*, 1993, 29(7): 2215–2226

Cote J, Konrad J M. A generalized thermal conductivity model for soils and construction materials. *Canadian Geotechnical Journal*, 2005, 42(2): 443–458

Covelo E, Vega F, Andrade M. Simultaneous sorption and desorption of Cd, Cr, Cu, Ni, Pb, and Zn in acid soils I. selectivity sequences. *Journal of Hazardous Materials*, 2007, 147: 852–861

Cui X, Wu D, Wang H, Ding S, Fan Y. Pore features and seepage characteristics of natural gap graded sand with two size distributions. *Géotechnique*, 2022, 73: 1003–1014

Cui X Z, Liu Q S, Zhang C Y. Detachment characteristics of deposited particles in porous medium: experimentation and modeling. *Transport in Porous Media*, 2017, 119(3): 633–647

Cundall P A, Strack O. A discrete numerical model for granular assemblies. *Geotechnique*, 1979, 29: 47–65.

Daghooghi M, Borazjani I. The effects of irregular shape on the particle stress of dilute suspensions. *Journal of Fluid Mechanics*, 2018, 839: 663–692

Deliyanni E A, Kyzas G Z, Triantafyllidis K S, Matis K A. Activated carbons for the removal of heavy metal ions: a systematic review of recent literature focused on lead and arsenic ions. *Open Chemistry*, 2015, 13(1): 699–708

Ding Y, Zhang J S, Jia Y, Chen X B, Meng F. Study on two-phase fluid-solid coupling characteristics in local-saturated zone of subgrade considering the effect of fine particles migration. *Applied Sciences*, 2020, 10(21): 7539

Du H, Huang Q, Peacock C L, Tie B, Lei M, Liu X, Wei X. Competitive binding of Cd, Ni and Cu on goethite organo-mineral composites made with soil bacteria. *Environmental Pollution*, 2018, 243: 444–452

Fang J, Owens R G, Tacher L, Parriaux P. A numerical study of the SPH method for simulating transient viscoelastic free surface flows. *Journal of Non-Newtonian Fluid Mechanics*, 2006, 139: 68–84

Farajzadeh R, Bedrikovetsky P, Lotfallahi M, Lake L W. Simultaneous sorption and mechanical entrapment during polymer flow through porous media. *Water Resources Research*, 2016, 52: 2279–2298

Feng Y, Zhao J, Hayati H, Sperry T, Yi H. Tutorial: Understanding the transport, deposition, and translocation of particles in human respiratory systems using computational fluid-particle dynamics and physiologically based toxicokinetic models. *Journal of Aerosol Science*, 2021, 151: 105672

Fourtakas G. *Modelling Multi-Phase Flows in Nuclear Decommissioning using SPH*. The University of Manchester, United Kingdom, 2015

Fox P J, Lee J, Lenhart J. Coupled consolidation and contaminant transport in compressible porous media. *International Journal of Geomechanics*, 2011, 11(2): 113–123

Fu Q, Du Z, Wang S. The study on the distribution law of interwell pressure gradient considering the influence of profile control and water shutoff. *Journal of Porous Media*, 2018, 21(3): 241–256

Funk M, Wakili K G. Driving potentials of heat and mass transport in porous building materials: a comparison between general linear, thermodynamics and micromechanical derivation schemes. *Transport in Porous Media*, 2008, 72: 273–294

Gan Z, Wang Z, Zhang Y, Zhao W, Li B. Pore-scale lattice Boltzmann model for heat and mass transfers in frozen soil. *International Journal of Heat and Fluid Flow*, 2024, 110: 109634

García J R, Bidabehere C María, Sedran U. Diffusion controlled LHHW kinetics. Simultaneous determination of chemical kinetic and equilibrium adsorption constants by using the Weisz-Prater approach. *Chemical Engineering Science*, 2017, 172: 444–452

Gavrilov A A, Shebelev A V. Single-fluid model of a mixture for laminar flows of highly concentrated suspensions. *Fluid Dynamics*, 2018, 53(2): 255–269

Gong H, Tan Z, Huang K, Zhou Y, Yu J, Huang Q. Mechanism of cadmium removal from soil by silicate composite biochar and its recycling. *Journal of Hazardous Materials*, 2021, 409: 125022

Grolimund D, Borkovec M, Barmettler K, Sticher H. Colloid-facilitated transport of strongly sorbing contaminants in natural porous media: a laboratory column study. *Environmental Science and Technology*, 1996, 30(10): 3118–3123

Guerrero J S P, Skaggs T H. Analytical solution for one-dimensional advection-dispersion transport equation with distance-dependent coefficients. *Journal of Hydrology*, 2010, 390(1–2): 57–65

Haliena B, Zheng H, Melson N, Kaplan D I, Barnett M O. Decreased salinity and actinide mobility: colloid-facilitated transport or pH change. *Environmental Science and Technology*, 2016, 50(2): 625–632

Hassini L, Raja L, Lecompte-Nana G L, Elcafsi M A. 2-D hydro-viscoelastic model for convective drying of deformable and unsaturated porous material. *Comptes Rendus Mecanique*, 2017, 345: 248–258

Hudcova B, Veselska V, Filip J, Cihalova S, Komarek M. Highly effective Zn(II) and Pb(II) removal from aqueous solutions using Mg-Fe layered double hydroxides: comprehensive adsorption modeling coupled with solid state analyses. *Journal of Cleaner Production*, 2018, 171: 944–953

Hyunjung N K, Sharon L W. Escherichia coli transport in porous media: influence of cell strain, solution chemistry, and temperature. *Colloids and Surface B-Biointerfaces*, 2009, 71: 160–167

Jahangir M H, Ghazvini M, Pourfayaz F, Ahmadi M H, Sharifpur M, Meyer J P. Numerical investigation into mutual effects of soil thermal and isothermal properties on heat and moisture transfer in unsaturated soil applied as thermal storage system. *Numerical Heat Transfer, Part A: Applications*, 2018, 73(7): 466–481

Jaiswal S, Chopra M, Das S. Numerical solution of a space fractional order solute transport system. *Journal of Porous Media*, 2018, 21(2): 145–160

James S C, Chrysikopoulos C V. Monodisperse and polydisperse colloid transport in water-saturated fractures with various orientations: gravity effects. *Advance in Water Resources*, 2011, 34: 1249–1255

Jeong J H, Jhon M S, Halow J S, Osdol J V. Smoothed particle hydrodynamics: applications to heat conduction. *Computer Physics Communications*, 2003, 153(1): 71–84

Jungnickel C, Smith D, Fityus S. Coupled multi-ion electrodiffusion analysis for clay soils. *Canadian Geotechnical Journal*, 2004, 41(2): 287–98

Karathanasis A D. Subsurface migration of copper and zinc mediated by soil colloids. *Soil Science Society of America Journal*, 1999, 63(4): 830–838

Kasper R, Turnow J, Kornev N. Numerical modeling and simulation of particulate fouling of structured heat transfer surfaces using a multiphase Euler-Lagrange approach. *International Journal of Heat and Mass Transfer*, 2017, 115: 932–945

Katzourakis V E, Chrysikopoulos C V. Mathematical modeling of colloid and virus cotransport in porous media: application to experimental data. *Advances in Water Resources*, 2014, 68: 62–73

Katzourakis V E, Chrysikopoulos C V. Impact of spatially variable collision efficiency on the transport of biocolloids in geochemically heterogeneous porous media. *Water Resources Research*, 2018, 54(6): 3841–3862

Katzourakis V E, Chrysikopoulos C V. Aggregating nanoparticle transport with nonlinear attachment: modeling and experimental validation. *Advances in Water Resources*, 2024, 193: 104819

Keller A A, Sirivithayapakorn S S, Chrysikopoulos C V. Early breakthrough of colloids and bacteriophage MS2 in a watersaturated sand column. *Water Resources Research*, 2004, 40(40): 474–480

Kersting A B, Efurd D W, Finnegan D L, Rokop D J, Smith D K, Thompson J L. Migration of plutonium in ground water at the Nevada Test Site. *Nature*, 1999, 397: 56–59

Kim H N, Walker S L. Escherichia coli transport in porous media: influence of cell strain, solution chemistry, and temperature. *Colloids and Surfaces B: Biointerfaces*, 2009, 71(1): 160–167

Kobayashi M, Nitanai M, Satta N, Adachi Y. Coagulation and charging of latex particles in the presence of imogolite. *Colloids and Surfaces A*, 2013, 435: 139–146

Kumar A, Jaiswal D K, Kumar N. Analytical solutions to one-dimensional advection-diffusion equation with variable coefficients in semi-infinite media. *Journal of Hydrology*, 2010, 380(3–4): 330–337

Kumpiene J. Management with willow short rotation coppice increase the functional gene diversity and functional activity of a heavy metal polluted soil. *Chemosphere*, 2015, 138: 469–477

Kumpiene J, Lagerkvist A, Maurice C. Stabilization of As, Cr, Cu, Pb and Zn in soil using amendments-a review. *Waste Management*, 2008, 28(1): 215–225

Lee S, Ko I W, Yoon I H, Kim D W, Kim K W. Colloid mobilization and heavy metal transport in the sampling of soil solution from Duckum soil in South Korea. *Environmental Geochemistry and Health*, 2019, 41(1): 469–480

Li C, Ran Y, Wu P, Liu P, Yang B, Gu X, Zhao P, Liu S, Song L, Liu Y, Liu Y, Ning Z, Sun J, Liu C. Antimony and arsenic migration in a heterogeneous subsurface at an abandoned antimony smelter under rainfall. *Journal of Hazardous Materials*, 2024, 470: 134156

Li G, Li X, Li C. Measurement of permeability and verification of Kozeny-Carman equation using statistic method. *Energy Procedia*, 2017, 142: 4104–4109

Li Y, Ahmadi A, Omari A, Pu H. Three-dimensional microscale simulation of colloidal particle transport and deposition in model porous media with converging/diverging geometries. *Colloids and Surfaces A*, 2018, 544: 179–186

Li Y C, Cleall P J. Analytical solutions for contaminant diffusion in double-layered porous media. *Journal of Geotechnical and Geoenvironmental Engineering*, 2010, 136(11): 1542–1554

Li Z L, Zhou L X. Cadmium transport mediated by soil colloid and dissolved organic matter: a field study. *Journal of Environmental Sciences*, 2010, 22(1): 106–115

Lindner L, Pfitzner M, Mundt C. Experimental and numerical study of multiphase, multicomponent flow in porous media with a multiphase mixture model. *Transport in Porous Media*, 2017, 116(1): 143-161

Liu M B, Liu G R. Smoothed particle hydrodynamics (SPH): An overview and recent developments. *Archives of Computational Methods in Engineering*, 2010, 17(1): 25–76

Liu Q, Cui X, Zhang C, Huang S. Experimental investigation of suspended particles transport through porous media: particle and grain size effect. *Environmental Technology*, 2016, 37(7): 854–864

Luo Z, Zhu J, Yu L, Yin K. Heavy metal remediation by nano zero-valent iron in the presence of microplastics in groundwater: inhibition and induced promotion on aging effects. *Environmental Pollution*, 2021, 287: 117628

Ma J, Guo H, Lei M, Wan X, Zhang H, Feng X, Wei R, Tian L, Han X. Blocking effect of colloids on arsenate adsorption during co-transport through saturated sand columns. *Environmental Pollution*, 2016, 213: 638–647

Malgaresi G, Khazali N, Bedrikovetsky P. Non-monotonic retention profiles during axi-symmetric colloidal flows. *Journal of Hydrology*, 2020, 580: 124235

Markale I, Velasquez-Parra A, Alcolea A, Jimenez-Martinez J. Mixing controlled adsorption at the liquid-solid interfaces in unsaturated porous media. *Transport in Porous Media*, 2022, 146(1–2): 159–175

Milosevic D, Levic S, Lazarevic S, Velickovic Z. Hybrid material based on subgleba of mosaic puffball mushroom (Handkea utriformis) as an adsorbent for heavy metal removal from aqueous solutions. *Journal of Environmental Management*, 2021, 297: 113358

Miracapillo C, Ferroni L. Numerical simulation of radionuclides migration in the far field of a geological repository. *Energy Procedia*, 2015, 82: 848–854

Missana T, Alonso U, Garcia-Gutierrez M. Role of bentonite colloids on europium and plutonium migration in a granite fracture. *Applied Geochemistry*, 2008, 23(6): 1484–1497

Monged M H E, Hussein M T, Khater A E M. Elemental and radiological aspects of geothermal springs and nearby soil and sediment of Al-Lith area: concentration and risk assessment. *Environmental Earth Sciences*, 2018, 77(427): 2–16

Moradi A, Smits K M, Lu N, McCartney J S. Heat transfer in unsaturated soil with application to borehole thermal energy storage. *Vadose Zone Journal*, 2016, 15(10): 1–13

Muraleedharan M, Nadir Y. Factors affecting the mechanical properties and microstructure of geopolymers from red mud and granite waste powder: a review. *Ceramics International*, 2021, 47(10): 13257–13279

Natarajan N, Kumar G S. Spatial moment analysis of colloid facilitated radionuclide transport in a coupled fracture-matrix system. *International Journal of Energy and Environment*, 2011, 2(3): 491–504

Nishad S, Al-Raoush R I. Colloid retention and mobilization mechanisms under different physicochemical conditions in porous media: a micromodel study. *Powder Technology*, 2021, 377: 163–173

Ovaysi S, Piri M. Pore-scale modeling of dispersion in disordered porous media. *Journal of Contaminant Hydrology*, 2011, 124(1–4): 68–81

Pang L, Close M E, Noonan M, Flintoft M, van den Brink P. A laboratory study of bacteria-facilitated cadmium transport in alluvial gravel aquifer media. *Journal of Environmental Quality*, 2005, 34: 237–247

Peld M, Tõnsuaadu K, Bender V. Sorption and desorption of Cd^{2+} and Zn^{2+} ions in apatite-aqueous systems. *Environmental Science and Technology*, 2004, 38: 256–277

Peters G P, Smith D W. Solute transport through a deforming porous medium. *International Journal for Numerical and Analytical Methods in Geomechanics*, 2002, 26(7): 683–717

Philip J R, De Vries D A. Moisture movement in porous materials under temperature gradients. *Transactions, American Geophysical Union*, 1957, 38(2): 222–232

Puls R W, Powell R M. Transport of inorganic colloids through natural aquifer material: implications for contaminant transport. *Environmental Science and Technology*, 1992, 26(3): 614–621

Rao D, Bai B. Study of the factors influencing diffusive tortuosity based on pore--scale SPH simulation of granular soil-. *Transport in Porous Media*, 2020, 132: 333–353

Rosenbrand E, Fabricius I L, Yuan H. Thermally induced permeability reduction due to particle migration in sandstones: the effect of temperature on kaolinite mobilization and aggregation. *Proceedings, Thirty-Seventh Workshop on Geothermal Reservoir Engineering*, 2012, SGP-TP-194

Russell T, Bedrikovetsky P. Colloidal-suspension flows with delayed fines detachment: analytical model and laboratory study. *Chemical Engineering Science*, 2018, 190: 98–109

Saavedra M, Miñarro M D, Angosto J M, Fernández-López J A. Reuse potential of residues of artichoke (Cynara scolymus L.) from industrial canning processin gas sorbent of heavy metals in multimetallic effluents. *Industrial Crops and Products*, 2019, 141: 111751

Saeibehrouzi L, Abolfathi S, Denissenko P, Holtzman R. Pore-scale modeling of solute transport in partially-saturated porous media. *Earth-Science Reviews*, 2024, 256: 104870

Sandra G, Mats J, Susanna W. Temperature effect on the stability of bentonite colloids in water. *Journal of Colloid and Interface Science*, 2006, 298: 694–705

Schembre J M, Kovscek. A R. Mechanism of formation damage at elevated temperature. *Journal of Energy Resources Technology*, 2005, 127(3): 171–180

Schwaiger H F. An implicit corrected SPH formulation for thermal diffusion with linear free surface boundary conditions. *International Journal for Numerical Methods in Engineering*, 2008, 75(6): 647–671

Seebergh J E, Berg J C. Evidence of a hairy layer at the surface of polystyrene latex particles. *Colloids and Surfaces A: Physicochemical Engineering Aspects*, 1995, 100: 139–153

Seetharam S C, Thomas H R, Vardon P J. Nonisothermal multicomponent reactive transport model for unsaturated soil. *International Journal of Geomechanics*, 2011, 11(2): 84–89

Sen T K, Khilar K C. Review on subsurface colloid and colloid-associated contaminant transport in saturated porous media. *Advances in Colloid and Interface Science*, 2006, 119(2–3): 71–96

Sendula A, Wang K, B. Subramanyam S, Cray J, Oram M, Chen X, Heitor A, Harbottle D, Brown K. Multiphase thermo-hydro-mechanical coupled soil drying model with phase-exchange based on mixture coupling theory. *International Journal of Engineering Science*, 2024, 203: 104119

Sharma K, Mallick S S, Mittal A. A study of energy loss due to particle and wall collisions during fluidized dense-phase pneumatic transport. *Powder Technology*, 2020, 362: 707–716

Shirvani M, Kalbasi M, Shariatmadari H, Nourbakhsh F, Najafi. Sorption-desorption of cadmium in aqueous palygorskite, sepiolite, and calcite suspensions: isotherm hysteresis. *Chemosphere*, 2006, 65: 2178–2184

Shu X, Wu Y, Zhang X, Yu F. Experiments and models for contaminant transport in unsaturated and saturated porous media—a review. *Chemical Engineering Research and Design*, 2023, 192: 606–621

Sim Y, Chrysikopoulos C V. Three dimensional analytical models for virus transport in saturated porous media. *Transport in Porous Media*, 1998, 30(1): 87–112

Sotirelis N P, Chrysikopoulos C V. Heteroaggregation of graphene oxide nanoparticles and kaolinite colloids. *Science of the Total Environment*, 2017, 579: 736–744

Sun H M, Gao B, Tian Y, Yin X Q, Yu C R, Wang Y Q, Ma L Q. Kaolinite and lead in saturated porous media: facilitated and impeded transport. *Journal of Environmental Engineering*, 2010, 136(11): 1305–1308

Syngouna V I, Giannadakis G I, Chrysikopoulos C V. Interaction of graphene oxide nanoparticles with quartz sand and montmorillonite colloids. *Environmental Technology*, 2020, 41(9): 1127–1138

Tang A M, Cui Y J. Effects of mineralogy on thermo-hydro-mechanical parameters of MX80 bentonite. *Journal of Rock Mechanics and Geotechnical Engineering*, 2010, 2(1): 91–96

Tang J, Zhu W, Kookana R, Katayama A. Characteristics of biochar and its application in remediation of contaminated soil. *Journal of Bioscience and Bioengineering*, 2013, 116(6): 653–659

Tari G, Olhero S M, Ferreira J M F. Influence of temperature on stability of electrostatically stabilized alumina suspensions. *Journal of Colloid and Interface Science*, 2000, 231: 221–227

Thomas H R, King S D. Coupled heat and mass transfer in unsaturated soil – A potential-based solution. *International Journal for Numerical and Analytical Methods in Geomechanics*, 1992, 16: 757–773

Thomas H R, He Y, Onofrei C. An examination of the validation of a model of the hydro/thermo/mechanical behavior of engineered clay barriers. *International Journal for Numerical and Analytical Methods in Geomechanics*, 1998, 22: 49–71

Tomita Y, Guo L C, Zhang Y, Uchida N, Uematsu K. Effect of temperature on the slurry characteristics and green bodies of alumina. *Journal of the American Ceramic Society*, 1995, 78(8): 2153–2156

Turkyilmazoglu M. Cooling of particulate solids and fluid in a moving bed heat exchanger. *Journal of Heat Transfer*, 2019, 141: 114501

Turkyilmazoglu M. Single phase nanofluids in fluid mechanics and their hydrodynamic linear stability analysis. *Computer Methods and Programs in Biomedicine*, 2020, 187: 105171

Tusara L, Itoi R, Yamashiro R, Fukuda D, Kawahara Y. Effects of suspended material and solution pH on solid deposition in porous media. *Journal of the Geothermal Research Society of Japan*, 2015, 37(4): 143–152

Valsala R, Govindarajan S K. Co-colloidal BTEX and microbial transport in a saturated porous system: numerical modeling and sensitivity analysis. *Transport in Porous Media*, 2019, 127: 269–294

Van Genuchten M T, Wierenga P J. Mass transfer studies in sorbing porous media: III. Experimental evaluation with 2,4,5-T. *Soil Science Society of America Journal*, 1977, 41(2): 276–285

Varnosfaderani A M, Gatmiri B, Haghighi E M. A model for moisture and heat flow in fractured unsaturated porous. *International Journal for Numerical and Analytical Methods in Geomechanics*, 2016, 41: 828–858

Vasiliadou I A, Chrysikopoulos C V. Cotransport of Pseudomonas putida and kaolinite particles through water saturated columns packed with glass beads. *Water Resources Research*, 2011, 47(2): 2144–2150

Wang H Q, Masssei N, Duont J P. Particle transport in porous medium: determination of hydrodispersive characteristics and deposition rates. *Earth and Planetary Sciences*, 2000, 331: 97–104

Wang J, Hu J, Zhang S. Studies on the sorption of tetracycline onto clays and marine sediment from seawater. *Journal of Colloid and Interface Science*, 2010, 349(2): 578–582

Wang K, Zheng L, Xie X, Chen X, Yu H. Unsaturated hydro-mechanical-electro-chemical coupling based on mixture-coupling theory: a unified model. *International Journal of Engineering Science*, 2023, 191: 103914

Wang Q, Cheng T, Wu Y. Distinct roles of illite colloid and humic acid in mediating Arsenate transport in water-saturated sand columns. *Water Air and Soil Pollution*, 2015, 226(5): 1–15

Wang Z, Fu S, Zhao X, Jiang W, Liu G, Li D, Qi H. Modeling temperature change and water migration of unsaturated soils under seasonal freezing-thawing conditions. *Cold Regions Science and Technology*, 2024, 218: 104074

Wang Z, Zhang W, Li S, Zhou J, Liu D. Transport of silica colloid through saturated porous media under different hydrogeochemical and hydrodynamic conditions considering managed aquifer recharge. *Water*, 2016, 8(12): 555

Wu G, Li L Y. Modeling of heavy metal migration in sand/bentonite and the leachate pH effect. *Journal of Contaminant Hydrology*, 1998, 33: 313–336

Wu H, Bai B, Nie Q, Jia X. The coupled migration model of two-phase flow with nonlinear adsorption-desorption patterns exhibiting delays. *International Journal of Heat and Mass Transfer*, 2025, 243: 126897

Wyness A J, Paterson D M, Defew E C, Stutter M I, Avery L M. The role of zeta potential in the adhesion of E. coli to suspended intertidal sediments. *Water Research*, 2018, 142: 159–166

Xiao Z, Li K, Duan J, Zhang S. Study on the multi-field-coupling model of saline frozen soil considering ice and salt crystallization. *Computers and Geotechnics*, 2024, 169: 106209

Xu J, Hu C, Wang M, Zhao Z, Zhao X, Cao L, Lu Y, Cai X. Changeable effects of coexisting heavy metals on transfer of cadmium from soils to wheat grains. *Journal of Hazardous Materials*, 2022, 423: 127182

Yadav S K, Kumar A, Jaiswal D K, Kumar N. One-dimensional unsteady solute transport along unsteady flow through inhomogeneous medium. *Journal of Earth System Science*, 2011, 120(2): 205–213

Yan H, Sedighi M, Ding H, Sun Z, Xie H. Analytical model for non-isothermal diffusion of contaminants in unsaturated composite liner. *Journal of Hydrology*, 2021, 603: 126848

Yin X, Gao B, Lena Q, Ma L Q, Saha U K, Sun H, Wang G. Colloid-facilitated Pb transport in two shooting-range soils in Florida. *Journal of Hazardous Materials*, 2010, 177: 620–625

Yu C, Liao R, Cai X, Yu X. Sodium polyacrylate modification method to improve the permeant performance of bentonite in chemical resistance. *Journal of Cleaner Production*, 2019, 213: 242–250

Zhang J, Chen Q, You C. Numerical simulation of mass and heat transfer between biochar and sandy soil. *International Journal of Heat and Mass Transfer*, 2015, 91: 119–126

Zhou D D, Jiang X H, Lu Y, Fan W, Hou M X, Crittenden J C. Cotransport of graphene oxide and Cu(II) through saturated porous media. *Science of the Total Environment*, 2016, 550: 717–726

Zhou F, Yang J, Ju W, Zhou Z, Ma Q. Study on the hydro-thermal-salt-mechanical coupling characteristics of sulfate saline soil under freeze-thaw cycles. *Cold Regions Science and Technology*, 2024, 226: 104289

Zhou L, Zhou F, Ying S, Li S. Study on water and salt migration and deformation properties of unsaturated saline soil under a temperature gradient considering salt adsorption: numerical simulation and experimental verification. *Computers and Geotechnics*, 2021, 134: 104094

Zhu Y, Ma L Q, Dong X, Yang W, Scheckel K A, Xie J. Ionic strength reduction and seepage interruption enhanced colloid-facilitated Hg transport in contaminated soils. *Journal of Hazardous Materials*, 2014, 264(2): 286–292

Zhu Z Q, Yang Y H, Fan Y M, Zhang L H, Tang S, Zhu Y N, Zhou X B. Strontium-doped hydroxyapatite as an efficient adsorbent for Cd(II) removal from wastewater: performance, kinetics, and mechanism. *Environmental Technology & Innovation*, 2022, 28: 102575

Ziedan W, Hussain W A, Ismail M M. Porous adsorbent based on kaolin and nanomagnetic cobalt ferrite for effective removal of Pb(II) from wastewater. *Journal of Superconductivity and Novel Magnetism*, 2024, 37: 587–596

Index

Note: Page numbers in *italics* indicate a figure and page numbers in **bold** indicate a table on the corresponding page.

D

E

F

G

H

I

L

M

N

P

Q

R

S

For Product Safety Concerns and Information please contact our EU
representative GPSR@taylorandfrancis.com
Taylor & Francis Verlag GmbH, Kaufingerstraße 24, 80331 München, Germany

www.ingramcontent.com/pod-product-compliance
Ingram Content Group UK Ltd.
Pitfield, Milton Keynes, MK11 3LW, UK
UKHW052221190926
475702UK00006B/1135

* 9 7 8 1 0 4 1 3 2 6 3 7 3 *